An Assessment of the Department of Energy's Office of Fusion Energy Sciences Program

Fusion Science Assessment Committee
Plasma Science Committee
Board on Physics and Astronomy
Division on Engineering and Physical Sciences
National Research Council

NATIONAL ACADEMY PRESS
Washington, D.C.

National Academy Press • 2101 Constitution Avenue, N.W. • Washington, DC 20418

NOTICE: The project that is the subject of this report was approved by the Governing Board of the National Research Council, whose members are drawn from the councils of the National Academy of Sciences, the National Academy of Engineering, and the Institute of Medicine. The members of the committee responsible for the report were chosen for their special competences and with regard for appropriate balance.

This project was supported by the Department of Energy under Grant No. DE-FG02-98ER54508.

International Standard Book Number 0-309-07345-6

Additional copies of this report are available from:

National Academy Press, 2101 Constitution Avenue, N.W., Lockbox 285, Washington, DC 20055; (800) 624-6242 or (202) 334-3313 (in the Washington metropolitan area); Internet <http://www.nap.edu>; and

Board on Physics and Astronomy, National Research Council, HA 562, 2101 Constitution Avenue, N.W., Washington, DC 20418; Internet <http://www.national-academies.org/bpa>.

Cover images: The three images on the cover illustrate the commonality of fundamental processes, such as for magnetic reconnection in astrophysical and laboratory plasmas. The top image, soft x-ray data from the TRACE (Transition Region and Coronal Explorer) satellite, shows loops of million-degree plasma in the solar corona. The heating of plasma in these loops is believed to result from the release of magnetic energy during magnetic reconnection. The four plots in the center of the cover are a sequence of electron temperature measurements from the core of the Tokamak Fusion Test Reactor during an "internal disruption." The topological change in the magnetic field resulting from magnetic reconnection facilitates the expulsion of the 50-million-degree plasma core. In the bottom image, the Magnetic Reconnection Experiment explores the basic physics of magnetic reconnection. Superimposed is a snapshot of the measured magnetic field. Courtesy of NASA and the Stanford-Lockheed Institute for Space Research (top) and M. Yamada, Princeton Plasma Physics Laboratory (middle and bottom).

Copyright 2001 by the National Academy of Sciences. All rights reserved.

Printed in the United States of America

THE NATIONAL ACADEMIES

National Academy of Sciences
National Academy of Engineering
Institute of Medicine
National Research Council

The **National Academy of Sciences** is a private, nonprofit, self-perpetuating society of distinguished scholars engaged in scientific and engineering research, dedicated to the furtherance of science and technology and to their use for the general welfare. Upon the authority of the charter granted to it by the Congress in 1863, the Academy has a mandate that requires it to advise the federal government on scientific and technical matters. Dr. Bruce M. Alberts is president of the National Academy of Sciences.

The **National Academy of Engineering** was established in 1964, under the charter of the National Academy of Sciences, as a parallel organization of outstanding engineers. It is autonomous in its administration and in the selection of its members, sharing with the National Academy of Sciences the responsibility for advising the federal government. The National Academy of Engineering also sponsors engineering programs aimed at meeting national needs, encourages education and research, and recognizes the superior achievements of engineers. Dr. William A. Wulf is president of the National Academy of Engineering.

The **Institute of Medicine** was established in 1970 by the National Academy of Sciences to secure the services of eminent members of appropriate professions in the examination of policy matters pertaining to the health of the public. The Institute acts under the responsibility given to the National Academy of Sciences by its congressional charter to be an adviser to the federal government and, upon its own initiative, to identify issues of medical care, research, and education. Dr. Kenneth I. Shine is president of the Institute of Medicine.

The **National Research Council** was organized by the National Academy of Sciences in 1916 to associate the broad community of science and technology with the Academy's purposes of furthering knowledge and advising the federal government. Functioning in accordance with general policies determined by the Academy, the Council has become the principal operating agency of both the National Academy of Sciences and the National Academy of Engineering in providing services to the government, the public, and the scientific and engineering communities. The Council is administered jointly by both Academies and the Institute of Medicine. Dr. Bruce M. Alberts and Dr. William A. Wulf are chairman and vice chairman, respectively, of the National Research Council.

FUSION SCIENCE ASSESSMENT COMMITTEE

CHARLES F. KENNEL, Scripps Institution of Oceanography, *Chair**
LINDA CAPUANO, Honeywell, Inc.
PATRICK L. COLESTOCK, Los Alamos National Laboratory
FRANCE CORDOVA, University of California at Santa Barbara*
JAMES F. DRAKE, University of Maryland*
NATHANIEL J. FISCH, Princeton University
LENNARD FISK, University of Michigan
RAYMOND FONCK, University of Wisconsin
ROBERT A. FROSCH, Harvard University*
GEORGE GLOECKLER, University of Maryland
ZORAN MIKIC, Science Applications International Corporation
ALBERT NARATH, Lockheed Martin Corporation (retired)*
CLAUDIO PELLEGRINI, University of California at Los Angeles*
STEWART C. PRAGER, University of Wisconsin at Madison*
ROBERT ROSNER, University of Chicago*
ANDREW M. SESSLER, Lawrence Berkeley Laboratory
ROBERT H. SOCOLOW, Princeton University*
JAMES W. VAN DAM, University of Texas at Austin
JONATHAN WURTELE, University of California at Berkeley

*Steering group member

DONALD C. SHAPERO, Board on Physics and Astronomy, Director
KEVIN D. AYLESWORTH, Program Officer (until October 15, 1999)
JOEL R. PARRIOTT, Program Officer
ACHILLES SPELIOTOPOULOS, Program Officer (after October 20, 2000)

PLASMA SCIENCE COMMITTEE

STEVEN C. COWLEY, University of California at Los Angeles, *Chair*
JILL P. DAHLBURG, Naval Research Laboratory
JAMES F. DRAKE, University of Maryland
MARK J. KUSHNER, University of Illinois at Urbana-Champaign
JOHN D. LINDL, Lawrence Livermore National Laboratory
DAVID MEYERHOFER, University of Rochester
THOMAS M. O'NEIL, University of California at San Diego
STEWART C. PRAGER, University of Wisconsin at Madison
ROBERT ROSNER, University of Chicago
JONATHAN WURTELE, University of California at Berkeley

DONALD C. SHAPERO, Board on Physics and Astronomy, Director
KEVIN D. AYLESWORTH, Program Officer (until October 15, 1999)
JOEL R. PARRIOTT, Program Officer (from October 16, 1999, to July 15, 2000)
ACHILLES SPELIOTOPOULOS, Program Officer (after July 15, 2000)

BOARD ON PHYSICS AND ASTRONOMY

JOHN P. HUCHRA, Harvard-Smithsonian Center for Astrophysics, *Chair*
ROBERT C. RICHARDSON, Cornell University, *Vice Chair*
GORDON A. BAYM, University of Illinois at Urbana-Champaign
WILLIAM BIALEK, NEC Research Institute
VAL FITCH, Princeton University
WENDY FREEDMAN, Carnegie Observatories
RICHARD D. HAZELTINE, University of Texas at Austin
KATHRYN LEVIN, University of Chicago
CHUAN LIU, University of Maryland
JOHN C. MATHER, NASA Goddard Space Flight Center
CHERRY ANN MURRAY, Lucent Technologies
JULIA PHILLIPS, Sandia National Laboratories
ANNEILA I. SARGENT, California Institute of Technology
JOSEPH H. TAYLOR, Jr., Princeton University
KATHLEEN TAYLOR, General Motors Research and Development Center
CARL E. WEIMAN, JILA/University of Colorado at Boulder
PETER G. WOLYNES, University of California at San Diego

DONALD C. SHAPERO, Director
ROBERT L. RIEMER, Program Officer
JOEL R. PARRIOTT, Program Officer
ACHILLES SPELIOTOPOULOS, Program Officer
SARAH CHOUDHURY, Senior Project Associate
NELSON QUIÑONES, Project Assistant

Preface

The Fusion Science Assessment Committee was convened under the auspices of the National Research Council's Board on Physics and Astronomy in 1998 in response to a request to the National Research Council by the Department of Energy's Office of Science for an assessment of the scientific quality of its Office of Fusion Energy Sciences (OFES) program.

The original charge to the committee was expanded in early 1999, after consultation with the associate director for OFES, to include a programmatic assessment in addition to the scientific assessment. The expanded charge to the committee is as follows: "The committee will assess the scientific quality of the fusion program of the DOE's Office of Science. Criteria will include excellence, impact, role in education, and contribution to strengthening the scientific foundation for fusion. A science strategy for the program will provide a context for judgement and a direction for future development." The committee was asked to look at OFES specifically, but in the course of its deliberations determined that it could not look at the OFES program in a meaningful way unless it considered the U.S. plasma science effort as a whole. Nonindustrial plasma science funding in this country is dominated by OFES, which now has a mandate to act as a steward for plasma science. Given this domination, the assessment refers to the OFES program and the overall U.S. plasma science research effort interchangeably. For the same reason, the committee uses the terms plasma science, plasma physics, and fusion science interchangeably in this report.

It is important to note topics that the committee did not directly address and that are therefore not included here. The report focuses on the science of magnetically confined plasmas and the programmatic strategy for long-term progress in this area, but it does not directly address inertially confined plasmas (with the exception of a brief mention in Chapter 3). The Department of Energy's Office of Defense Programs sponsors major inertial-confinement fusion research for stockpile stewardship purposes. OFES also sponsors a relatively small, and complementary, program in inertial fusion driver and target research relevant to the energy goal. Although some plasma science issues are common to both magnetic and inertial confinement, the programs are structured quite differently. Also, this assessment does not directly address issues surrounding specific technology development and engineering research sponsored by the program (with the exception of a brief mention in Chapter 3), because the committee chose to focus on elements of the program related to basic plasma physics research.

Although the committee explicitly mentions the International Toroidal Experimental Reactor (ITER) only in connection with the 1996 redirection of OFES, the existence of this international partnership was implicit during its deliberations, and international efforts are mentioned in connection with the potential for a future burning experiment.

The committee divided into four groups to facilitate the writing of this report, although the full committee participated in the discussions, data gathering, and internal review of the full report. The four groups were a steering group and three working groups, which wrote the three main chapters of the report. The steering group consisted of the chair, four senior science and technology policy experts, and the leaders of the three working groups. The working groups were led by James Drake, Stewart Prager, and Robert Rosner, and each was made up of members from both inside and outside fusion science. The insiders provided the expertise and had detailed knowledge of the history of the fusion program, and the outsiders acted as objective reviewers. Short summaries of the five committee meetings are found in Appendix A.

In September 1999, the committee produced an interim report[1] on the quality of the science in the OFES. OFES requested this interim report so it could meet its obligations under the Government Performance Review Act. The interim report was necessarily brief, lacked detailed findings, and was without any recommendations or discussion of programmatic issues. Despite such shortcomings, its findings serve as the core of the discussion in Chapter 4 of this final report.

To learn the views of the stakeholders in the federal government, the chair held discussions with representatives of the Department of Energy, the Congress, the Office of Management and Budget, the Office of Science and Technology Policy, and the Congressional Research Service. For an excellent in-depth analysis of the history of the U.S. fusion program, the committee strongly recommends *Congress and the Fusion Energy Sciences Program: A Historical Analysis.*[2]

The committee would like to acknowledge the technical input of Martin Greenwald, Alan Turnbull, Ted Strait, Tony Taylor, Frank Waelbroeck, Michael Zarnstorff, Cary Forest, Keith Burrell, Edward Synakowski, and William Dorland. It would also like to thank Kenneth Gentle for providing the results of his independent demographic study of plasma physics faculty members at universities. The committee would also like to acknowledge two speakers at the La Jolla meeting whose talks helped to lay the foundation for the science assessment found in this report: Predhiman Kaw, Institute for Plasma Research, India, on the international standing of the U.S. fusion science program, and Steven Cowley, University of California at Los Angeles, on contributions of fusion theory to nonfusion science. The committee acknowledges the excellent assistance of the NRC staff, including Don Shapero, Joel Parriott, and Kevin Aylesworth.

It has been a privilege and a pleasure for this committee, a blend of practicing fusion scientists, active scientists in related fields, and scientists with considerable managerial experience, to get to know fusion research at a critical turning point in its evolution. The committee hopes that the different perspectives of the members will stimulate the fusion research community to reach out to the broader science community and thereby to assume its rightful place in the sun.

Charles F. Kennel, *Chair*
Fusion Science Assessment Committee

---

[1]National Research Council, Fusion Science Assessment Committee. 1999. *Interim Report.* Washington, D.C.: National Academy Press.

[2]Richard E. Rowberg. 2000. *Congress and the Fusion Energy Sciences Program: A Historical Analysis.* Washington, D.C.: Congressional Research Service.

Acknowledgment of Reviewers

This report has been reviewed in draft form by individuals chosen for their diverse perspectives and technical expertise, in accordance with procedures approved by the NRC's Report Review Committee. The purpose of this independent review is to provide candid and critical comments that will assist the institution in making its published report as sound as possible and to ensure that the report meets institutional standards for objectivity, evidence, and responsiveness to the study charge. The review comments and draft manuscript remain confidential to protect the integrity of the deliberative process. We wish to thank the following individuals for their review of this report:

Abraham Bers, Massachusetts Institute of Technology,
Steven C. Cowley, University of California at Los Angeles,
Marvin L. Goldberger, University of California at San Diego,
William Happer, Jr., Princeton University,
Chuan Sheng Liu, University of Maryland,
Roy F. Schwitters, University of Texas at Austin,
Clifford M. Surko, University of California at San Diego,
Lilian Shiao-Yen Wu, IBM Corporation, and
Ellen G. Zweibel, University of Colorado.

Although the reviewers listed above provided many constructive comments and suggestions, they were not asked to endorse the conclusions or recommendations nor did they see the final draft of the report before its release. The review of this report was overseen by John F. Ahearne, of Sigma Xi, the Scientific Research Society, and Duke University, appointed by the NRC's Report Review Committee, who was responsible for making certain that an independent examination of this report was carried out in accordance with institutional procedures and that all review comments were carefully considered. Responsibility for the final content of this report rests entirely with the authoring committee and the institution.

Contents

Executive Summary   1

1 OVERVIEW: ASSESSMENT AND HISTORICAL CONTEXT   9
Assessment of Quality: Scientific Progress and the Development of Predictive
      Capability, 9
Program Development: Plasma Confinement Configurations, 11
Institutional Considerations: Interactions of the Fusion Program With Allied
      Areas of Science and Technology, 12

**2 SCIENTIFIC PROGRESS AND THE DEVELOPMENT OF PREDICTIVE
      CAPABILITY**   14
Summary, 15
    Equilibrium and Heating, 15
    Stability, 16
    Transport, 16
**Equilibrium and Heating: Designing, Controlling, and Diagnosing a Confined
      Plasma,** 17
    Tools for Calculating Equilibria, 18
    Tools for Measuring Equilibrium Pressure and Magnetic and Electric Fields, 19
    Tools for Controlling Plasma Properties, 20
Stability: Predicting Operational Behavior, 23
    Ideal Stability of Confined Plasmas, 24
    Nonideal Instabilities and Magnetic Reconnection, 26
    Density Limits, 29
    Influence of Fast Particles, 30

Transport: Ensuring Sufficient Confinement, 32
    Empirical Scaling Law Approach, 33
    Development of Tools for Calculating Stability and Simulating Nonlinear Microturbulence, 34
    Development of Tools for Remote Measurement of Fluctuations and Transient Phenomena, 36
    Transport Barriers and Confinement Control, 37
    Evaluation of the Present Understanding of Turbulent Transport, 39
Findings and Recommendations, 42
    Findings, 42
    Recommendations, 43

3 PLASMA CONFINEMENT CONFIGURATIONS   45
Introduction, 45
Important Physics Questions Motivating Research With Various Configurations, 46
    Understand the Stability Limits to Plasma Pressure, 46
    Understand and Control Magnetic Chaos in Self-Organized Systems, 48
    Understand Classical Plasma Behavior and Magnetic Field Symmetry, 49
    Understand Plasmas Self-Sustained by Fusion ("Burning" Plasmas), 51
Reactor Design Features Motivating Fusion Concept Development, 53
Inertial Fusion Energy Concept Development, 55
Linkages With International Programs, 56
Enabling Technologies for Plasma Configuration Development, 56
Current Metrics for Fusion Concept Development, 57
Findings and Recommendations, 60
    Findings, 60
    Recommendations, 60

4 INTERACTIONS OF THE FUSION PROGRAM WITH ALLIED AREAS OF SCIENCE AND TECHNOLOGY   62
Introduction, 62
What Are the Deep Scientific Contributions That Have Impacted Other Physics Fields?, 63
    Stability Theory, 63
    Stochasticity, Chaos, and Nonlinear Dynamics, 63
    Dissipation of Magnetic Fields, 63
    Origins of Magnetic Fields, 64
    Wave Dynamics, 64
    Turbulent Transport, 65
What Have Been the Fusion-specific Contributions of the United States to the World Program?, 65
What Are the Future Forefront Areas of Interdisciplinary Research?, 66
Does the Field Maintain Leadership in Key Supporting Research Areas?, 71
    Computational Physics, 72
    Applied Mathematics, 74
    Experimental Techniques, 75

Has the Field Been Recognized for Asking and Answering Deep Physics Questions?, 76
Are Plasma Scientists Well Represented and Trained at the Nation's Major Research Universities?, 76
Creation of Centers of Excellence in Plasma and Fusion Science, 77
Findings and Recommendations, 79
    Findings, 79
    Recommendations, 79

APPENDIXES   83
A **Summary of Committee Meetings**, 85
B **Funding Data**, 87
    Total Budget of the Office of Fusion Energy Sciences, 87
    National Science Foundation/Department of Energy Plasma Physics Partnership Funding, 89
    Office of Fusion Energy Sciences Funding to Universities, 89
C **The Family of Magnetic Confinement Configurations**, 91
    The Stellarator, 91
    The Tokamak, 92
    The Spherical Torus, 92
    The Reversed-Field Pinch, 92
    The Spheromak and the Field-Reversed Configuration, 92
    Other Concepts, 93
D **Glossary**, 94
E **Acronyms and Abbreviations**, 96

Executive Summary

The purpose of this assessment of the fusion energy sciences program of the Department of Energy's (DOE's) Office of Science is to evaluate the quality of the research program and to provide guidance for the future program strategy aimed at strengthening the research component of the program.[1] The committee focused its review of the fusion program on magnetic confinement, or magnetic fusion energy (MFE), and touched only briefly on inertial fusion energy (IFE), because MFE-relevant research accounts for roughly 95 percent of the funding in the Office of Science's fusion program. Unless otherwise noted, all references to fusion in this report should be assumed to refer to magnetic fusion.

Fusion research carried out in the United States under the sponsorship of the Office of Fusion Energy Sciences (OFES) has made remarkable strides over the years and recently passed several important milestones. For example, weakly burning plasmas with temperatures greatly exceeding those on the surface of the Sun have been created and diagnosed. Significant progress has been made in understanding and controlling instabilities and turbulence in plasma fusion experiments, thereby facilitating improved plasma confinement—remotely controlling turbulence in a 100-million-degree medium is a premier scientific achievement by any measure. Theory and modeling are now able to provide useful insights into instabilities and to guide experiments. Experiments and associated diagnostics are now able to extract enough information about the processes occurring in high-temperature plasmas to guide further developments in theory and modeling. Many of the major experimental and theoretical tools that have been developed are now converging to produce a qualitative change in the program's approach to scientific discovery.

The U.S. program has traditionally been an important source of innovation and discovery for the international fusion energy effort. The goal of understanding at a fundamental level the physical processes governing observed plasma behavior has been a distinguishing feature of the program. This feature, a strength of the program, was formalized in the 1996 restructuring, with the new emphasis on

---

[1] Because OFES funding so dominates the funding of plasma science in the United States, this report uses the terms "fusion science," "plasma physics," "plasma science," and "the fusion program" interchangeably.

establishing the knowledge base for fusion energy.[2] An essential tool to unravel the complexities of plasma dynamics is a strong theory program. For several decades, the United States has played a dominant role in plasma theory. The quantitative detail in which experiments are designed and executed in the United States has become a benchmark for the rest of the world. However, the close interaction between the U.S. and international programs since the 1950s makes it difficult to separate the U.S. contributions from those of other countries.

Mutual reinforcement of theory and experiment, strong international leadership, and discovery of fundamental principles are hallmarks of a successful scientific enterprise. The committee concludes, therefore, that *the quality of the science funded by the United States fusion research program in pursuit of a practical source of power from fusion (the fusion energy goal) is easily on a par with the quality in other leading areas of contemporary physical science.*

However, in spite of the high quality of the science being carried out, some serious demographic and sociological problems—caused in part by programmatic emphasis and in part by organizational structures—must be addressed. As outlined in the interim report[3] of the committee, there is a history of intellectual interchange between the fusion plasma community and the broader scientific community. Nevertheless, the increasing focus on the fusion energy goal prior to 1996 gradually caused the fusion program to become too inward looking and therefore intellectually isolated from the rest of science—fusion science was not seen as a generator of ideas impacting other scientific disciplines. While many of the scientific challenges that must be overcome in pursuit of the energy goal are sufficiently important to have a potentially broad impact on other branches of science, most scientists funded by the program do not actively participate in the wider scientific culture. As a result, the flow of scientific information out of and into the field is weak. New ideas and techniques developed in allied fields are slow to percolate into the program. Nor is the high-quality science in the program widely appreciated outside the field. Indeed, the broader scientific community holds a generally negative view of fusion science. This isolation, combined with the generally negative perception of the field, is reducing the number of universities and laboratories where plasma and fusion science is being studied to a degree that endangers the future of plasma science. The proportion of the program based on open, competitive, peer-reviewed grants is small, which discourages the entry of new talent into the field and further increases the isolation.

The committee believes that a dynamic, outward-looking, science-driven program in which discoveries are regularly communicated beyond the walls of fusion science is essential to alter the outside community's perception of the field. A strong case can also be made that a program organized around critical science goals will also maximize progress toward a practical fusion power source. Scientific discoveries that a decade ago would have been unthinkable are the fundamental drivers of program direction at all levels (see the third finding in Chapter 2). Thus, scientific discovery is inherently coupled with progress toward fusion, and the two should not be considered opposing forces.

---

[2]In 1996, the goal of the fusion program as a schedule-driven energy-development program was altered to reflect a longer term strategy for developing and deploying fusion energy sources. The central goal of the restructured program is to establish the knowledge base needed for an economically and environmentally attractive fusion energy source. See Department of Energy (DOE), Fusion Energy Advisory Committee. 1996. *A Restructured Fusion Sciences Program.* Washington, D.C.: DOE.

[3]National Research Council, Fusion Science Assessment Committee. 1999. *Interim Report.* Washington, D.C.: National Academy Press.

PRIMARY RECOMMENDATIONS

The committee makes seven primary recommendations, which address important concerns about the future of the fusion science effort. The findings are summarized in this chapter and explained in full at the end of Chapters 2, 3, and 4. Those chapters also contain secondary recommendations, which are more specific than the primary recommendations.

Recommendation 1. Increasing our scientific understanding of fusion-relevant plasmas should become a central goal of the U.S. fusion energy program on a par with the goal of developing fusion energy technology, and decision making should reflect these dual and related goals.

Since the redirection of the fusion program in 1996, a greater emphasis has been placed on understanding the basic plasma dynamics underlying the operation of the various confinement configurations. The new emphasis on exploring scientific issues has been effectively implemented on individual experiments. However, at the programmatic level, performance goals rather than overarching scientific goals continue to act as the primary driver for the allocation of resources. (See Chapter 3 for further discussion.) This emphasis is reflected, for example, in the categorization of experiments as concept exploration, proof of principle, and performance extension, which appears to measure the reactor potential of an experiment rather than its scientific merit—there is no parallel measure of scientific worth. Given the significant historical impact of scientific discovery on the program, the absence of a science-based strategic planning process is inhibiting progress.

DOE, in full consultation with the scientific community, needs to define a limited set of important scientific goals for fusion energy science and should formulate concrete and specific strategies to achieve each goal. The committee understands that such a planning process is under way, but it is premature to make a judgement about this new endeavor.

The accomplishment of the scientific goals should serve as a metric of the program's success and should have the same weight as performance, which is now the primary measure of progress. Progress in our scientific understanding of fusion-relevant plasmas and progress toward fusion energy are coupled, and both should serve to assess the program. The program planning and budget justification carried out by DOE must be organized around answering key scientific questions in fusion-relevant plasmas as well as around progress toward the eventual energy goal. This recommendation applies to the confinement configuration program and to other programs of a more general nature.

Public and congressional advocacy should insist on progress in science as well as progress toward a practical source of fusion power.

Recommendation 2. A systematic effort to reduce the scientific isolation of the fusion research community from the rest of the scientific community is urgently needed.

Program planning, funding, and administration should encourage connectivity with the broad scientific community. The community of fusion scientists should make a special effort to communicate its concepts, methods, tools, and results to the wider world of science, which is largely unaware of that community's recent scientific accomplishments. Increased connectivity will also facilitate the transfer of new ideas and techniques into the program from allied fields, enhancing the ability of the program to maximize the rate of scientific discovery.

There are numerous examples in federally funded research programs of formal coordination mechanisms having been established among related programs in different agencies. In some instances this coordination can optimize the use of funding. Perhaps more significant, the dialogue among the leaders of these government research programs can encourage interactions among the various scientific communities, foster joint undertakings, and raise the visibility of the discipline as a whole.

Recommendation 3. The fusion science program should be broadened in terms of both its institutional base and its reach into the wider scientific community; it should also be open to evolution in its content and structure as it strengthens its research portfolio.

The committee is convinced that the opportunity to understand the plasma physics underlying fusion is expanding because of the closer connection between theory and experiment and the great improvements in diagnostics and numerical simulation. To enrich the pool of ideas that will feed this progress, it will be essential to enlarge the sphere of awareness of the critical problems facing the field and to bring in new talent, both individual and institutional.

The broadening of the fusion science effort can be approached in a number of ways. One approach would be to set up competitive funding opportunities of sufficient magnitude to elicit responses from potential new institutional participants. The creation of centers of excellence in fusion science (proposed below) and the greater involvement of the National Science Foundation in fusion and plasma science would also broaden the institutional base of fusion science.

A larger proportion of fusion funding should be made available through open, well-advertised, competitive, peer-reviewed solicitations for proposals. Fusion program peer review could involve scientists from outside the fusion community where appropriate. The evaluation and ranking of proposals by panels that include individuals with appropriate expertise in allied fields would broaden the intellectual reach of the grant review process.

Plasma science research that is not immediately related to the fusion energy goal should play a more influential role in the DOE fusion program. More plasma science could be included in DOE's fusion science portfolio by having a program element for general plasma science. This program element should award individual investigator grants on a competitive, peer-review basis. A small fraction of the present DOE program addresses this need (see Appendix B), but its role and visibility should be increased. Such funding would encourage new interchanges that enrich fusion science.

To ensure that increasing institutional diversity is a continuing goal, the committee recommends that the breadth and flexibility of participation in the fusion energy science program be a program metric.

Recommendation 4. Several new centers, selected through a competitive, peer-review process and devoted to exploring the frontiers of fusion science, are needed for both scientific and institutional reasons.

Many of the issues in fusion science are now of sufficient complexity that they require closely interacting, critical-mass groups of scientists to make progress. For example, understanding the dynamics of plasma turbulence and transport requires the development of appropriate physical models and computational algorithms for treating disparate space- and timescales, as well as complex magnetic geometries, efficient programming on massively parallel computing platforms, and an understanding of nonlinear physics (energy cascades, intermittency, phase transitions, avalanches) and other topics. Tight coupling with a parallel experimental effort is required to challenge theoretical predictions.

No single scientist and no small collaboration of practicing scientists has the breadth of knowledge required to tackle such large and complex problems. In the area of theory and computation, the absence of closely interacting teams of critical mass is inhibiting the successful attack on a number of central science issues confronting the fusion research program. The loose collaborations that have been periodically established by the program have generally not been successful in establishing the close working relationships required to address the most challenging topics.

The new centers ("centers of excellence") could create a new focus on scientific issues for the U.S. fusion program. A center could serve as a node for a distributed network of close collaborators or it could undertake scientific explorations of significant magnitude at one site, or it could do both. The centers could marry the expertise and approaches of national labs and universities around the country. They should have a number of programmatic and structural features so that they can play their appropriate role in addressing the critical problems of the field. Among these features should be the following:

- A proposal for a center should have a plan to identify, pose, and answer scientific questions whose importance is widely recognized.
- One size cannot meet all scientific challenges. The committee envisions a center comparable in size to the current centers sponsored by the National Science Foundation (NSF), which have operating costs of $1 million to $5 million per year, although the size should ultimately be determined by the proposal process. Some centers may need on-site experimental facilities, and some may need only computing facilities and access to larger national computer centers.
- A team of between four and six coinvestigators with broad expertise and connections to other research groups and laboratories should form the core of the center's personnel. This team should be augmented by a similar number of temporary research staff (funded, at least in part, by the center) and an appropriate number of support staff.
- The center should enable links to various scientific disciplines, including physics, mathematics, and computer science, depending on the problem it is focusing on. It should have a plan for bringing practitioners of other disciplines from other institutions into the fusion community and should make the community's experimental resources more widely available.
- The institutions housing or participating in such centers should make a commitment to add faculty or career staff, as appropriate, in plasma/fusion science and/or related areas.
- The centers should have a strong educational component, featuring outreach to local high schools, undergraduate research opportunities, and a graduate research program.
- Centers should sponsor multidisciplinary workshops and summer schools focused on their central problem that will bring together students and researchers from various fields and institutions. The workshops would aim to bring in new ideas and collaborators as well as to disseminate to other fields the results they are achieving as they address the fundamental problems of fusion science.

Potential focus topics for centers include turbulence and transport, magnetic reconnection, energetic particle dynamics, and materials; other topics would emerge in a widely advertised proposal process. Topics such as these are of broad scientific interest in allied fields. To build another bridge to allied fields, the DOE should cooperate with the NSF in establishing one or more centers addressing a topic of general interest in plasma science. The DOE/NSF centers should have as a central objective establishing collaborations with scientists who have expertise of value to the plasma science and fusion research effort. An explicit goal of the centers should be to convey important scientific results to the broader scientific community as well as the rest of the fusion community. An announcement of opportunity for fusion centers of excellence would, by itself, signal to the broader scientific community the community's intent to significantly bolster the scientific strength of the field.

It would be highly desirable for other agencies, particularly NSF, to collaborate in one or more fusion centers of excellence for reasons of disciplinary and institutional diversity as well as to obtain the benefits of interagency collaboration cited in recommendation 3. However, the DOE should play a lead role in these centers, not only for reasons of administrative clarity but also because its leadership will ensure that the impressive capabilities of the fusion energy science community are made available to new participants. DOE leadership will also ensure that progress in the centers would be communicated throughout the fusion community and translated into DOE program plans, to hasten progress towards the fusion energy goal.

The procedure for awarding grants for fusion centers of excellence could do much to remedy the isolation of the fusion science community by ensuring that the broader scientific community will participate in the institution-building effort. The selection process for the centers should feature open, competitive peer review employing clear, science-based selection criteria.

The committee believes that the establishment of such centers is critical enough to the new science-based approach to fusion energy that ways should be found to fund a first center *even in a level budget scenario*. The success of the competition and the quality of the first center should guide the decision on launching second or even third centers. In other programs, such centers have been effective mechanisms for broadening and deepening a scientific area. In the committee's view, there is a very strong argument for expanding program funding to give fusion centers of excellence a strong and durable foundation.

Recommendation 5. Solid support should be developed within the broad scientific community for U.S. investment in a fusion burning experiment.

A burning plasma experiment will eventually be scientifically necessary and is on the critical path to fusion energy. Determining the optimal route to a burning plasma experiment is beyond the scope of the committee; rather, the route should be decided in the near term by the fusion community. Resources above and beyond those for the present program will be required. The U.S. scientific community needs to take the lead in articulating the goals of an achievable, cost-effective scientific burning experiment and to develop flexible strategies to achieve it, including international collaboration.

The committee agrees with the Secretary of Energy Advisory Board (SEAB) report that "… development both of understanding of a significant new project and of solid support for it throughout the political system is essential."[4] However, since the U.S. fusion energy effort is now positioned strategically as a science program, advocacy by the larger scientific community for the next U.S. investments in a fusion burning experiment now becomes even more critical to developing that support. For this reason alone, the scientific isolation of the fusion science community needs to be addressed.

Recommendation 6. The National Science Foundation should play a role in extending the reach of fusion science and in sponsoring general plasma science.

The mission of OFES, following the restructuring of the program in 1996, has been to establish the knowledge base in plasma physics required for fusion energy, with the result that a substantial number of plasma science problems are being explored within the fusion regime that also have applicability to allied fields such as astrophysics. For this reason, the committee believes that the NSF should begin to play a larger role in the solution of these basic plasma science problems. The greater involvement of NSF could have an intellectual impact on basic plasma science similar to that which it has had on basic

---

[4]Department of Energy (DOE), Secretary of Energy Advisory Board (SEAB), Task Force on Fusion Energy. 1999. *Realizing the Promise of Fusion Energy: Final Report of the Task Force on Fusion Energy.* Washington, D.C.: DOE, p. 2.

research in other scientific disciplines where mission agencies like DOE play the main funding role. NSF involvement would facilitate linkage to other fields and the involvement of new scientists in the program.

Recently, NSF and DOE collaborated on a small but highly effective program to encourage small laboratory plasma experiments and the theoretical exploration of topics in general plasma science. The large number of proposals submitted to this program is an indication of the need for it. The rationale for the expansion of research in general plasma science was well articulated in an earlier document of the National Research Council (NRC).[5]

The NSF/DOE plasma science initiative, if operated at a dollar level closer to that contemplated in the *Plasma Science* report (an additional $15 million per year for basic experiments in plasma science), can serve several important functions:

- Stimulating research on broad issues in plasma science that have potential applications to fusion and
- Enhancing interagency cooperation and cultural exchange on the approaches used by the two agencies for defining program opportunities, disseminating information on research results to the scientific community, selecting awardees, and judging the outcomes of grants.

The optimal process for this partnership, if there is sufficient funding (as requested in the *Plasma Science* report), would be an annual solicitation of requests for proposals (RFPs). In particular, this frequency would give new Ph.D.s the chance to enter the field and stay there, since new Ph.D.s are produced by degree-granting institutions each year and new graduate students enter school each year.

Another limitation of the ongoing NSF/DOE program in basic plasma science is the absence of any provision for modest experiments in the $1-million-per-year class. Historically, neither DOE nor NSF has funded plasma science experiments of this scale. For this reason, the committee recommends a cooperative NSF/DOE effort to broaden the scientific and institutional reach of fusion and plasma research to obtain valuable scientific results. Increased NSF funding and a stronger focus on fusion and plasma science within NSF would be required. As discussed in recommendation 4, NSF could cosponsor one or more centers of excellence in fusion and plasma science.

Recommendation 7. There should be continuing broad assessments of the outlook for fusion energy and periodic external reviews of fusion energy science.

The committee finds the current pattern of multiple program reviews of different provenance to be excessive. A planned sequence of external reviews of the U.S. fusion science program should replace this pattern. The reviews should be open, independent, and independently managed. They should involve a cross section of scientists from inside and outside the fusion energy program. The manifest independence of the review process will help ensure the credibility of the reviews in the eyes of Congress, the Office of Management and Budget (OMB), and the broader scientific community.

The scientific, engineering, economic, and environmental outlook for fusion energy should be reviewed every 10 years or so in a process that draws on fusion scientists, other scientists, engineers, policy planners, environmental experts, economists, and others, from here and abroad. These reviews

---

[5] National Research Council, Panel on Opportunities in Plasma Science and Technology. 1995. *Plasma Science: From Fundamental Research to Technological Applications.* Washington, D.C.: National Academy Press.

should assess from multiple perspectives the progress in the critical interplay between fusion science and engineering in light of the evolving technological, economic, and social contexts for fusion energy.

Consonant with its charge, the committee has not taken up the many critical-path issues associated with basic technology development for fusion, nor has it looked at the engineering of fusion energy devices and power plants, yet it is the combined progress made in science and engineering that will determine the pace of advancement toward the energy goal. Moreover, since much of fusion science research is undertaken in the expectation that it will contribute to the energy goal, regular, formal assessment of the progress towards fusion energy is one important way in which a fusion science program can be made accountable.

STRUCTURE OF THE REPORT

The report is organized into an overview chapter and three working group chapters: "Scientific Progress and the Development of Predictive Capability," "Plasma Confinement Configurations," and "Interactions of the Fusion Program with Allied Areas of Science and Technology."

Chapter 1 summarizes the general findings of the committee that came out of the three working group chapters and committee deliberations. It also refers to other reviews of the program since 1996, touches on the recent history of the tokamak experimental effort, and mentions briefly international efforts to build a fusion reactor, the International Toroidal Experimental Reactor (ITER).

The first priority of the committee was an evaluation of the science being carried out by the DOE's OFES program. Chapter 2 of the report examines the science being done in the program, both in areas where there is state-of-the-art understanding and in areas where further work is needed to achieve the predictive capability that will facilitate the design of the optimum magnetic container for holding hot fusion-grade plasmas. Chapter 2 is the longest chapter of the report since the committee felt an in-depth elucidation of the physics was essential to a proper evaluation. Readers who want only a summary of the scientific progress may read the summary section of the chapter.

In Chapter 3 of the report the various classes of magnetic container or confinement configuration are discussed. The discussion of the devices is organized around four scientific topics to illustrate the commonality of the physics issues being explored in each class of container. This commonality is too often lost when the devices are discussed separately. Enabling technologies and metrics are also discussed. Appendix C presents paragraph-long descriptions of each of the major confinement concepts.

Chapter 4 addresses the interaction between the field of fusion and plasma science and the larger science, engineering, and technology community. Discussed are the deep physics questions that have been addressed by the program, including the generic science results that have impacted other areas of physics, some of the key contributions of U.S. scientists to the international effort, potentially important areas of future interdisciplinary research, leadership in the support of research areas, and recognition for scientific accomplishments.

In addition to Appendixes A through C, which were mentioned above, Appendix D is a glossary of technical terms and Appendix E lists acronyms and abbreviations.

1

Overview: Assessment and Historical Context

The quest for fusion energy is one of the greatest scientific and technological challenges ever attempted, driving the development of modern plasma science and of the technology needed to work with one of the most hostile experimental environments on Earth.

The committee, through its three working groups that produced the main chapters of this report, studied the science being carried out in the fusion program, the various past and current plasma confinement configurations, and the interaction of the fusion science program with allied areas of science and technology. This examination of the magnetic confinement plasma research funded by the U.S. fusion program and of the standing of this research in the international fusion community has led to assessments in three areas: scientific progress, the development of plasma confinement configurations, and links to the broader scientific community.

ASSESSMENT OF QUALITY: SCIENTIFIC PROGRESS AND THE DEVELOPMENT OF PREDICTIVE CAPABILITY

The fusion research community has produced a continuing stream of scientific discoveries and technological innovations in pursuit of a practical fusion power source (the fusion energy goal). The research problems central to contemporary magnetic fusion plasma science—turbulence and transport, the limits on the magnetic confinement of plasma energy density, and the physics of reacting, self-heated plasmas—are extraordinarily challenging. The challenge—physical probes cannot survive the intense thermal fluxes in the high-temperature regime of interest to fusion research—is a consequence of the strong nonlinearity of the plasma medium and its resulting complex and rich dynamics. The study of high-temperature plasmas motivated by the fusion goal has become a fundamental branch of physics and one that is of great intrinsic interest. The broader influence of fusion science is clear from its impact on other branches of science, including astrophysics, space plasma physics, nonlinear science, optics, and turbulence.

Fusion research recently passed several important milestones. Weakly burning plasmas were created and confined for the first time, indicating that many of the tools have now been assembled for answering the scientific questions related to the fusion energy goal. The basic understanding of some of the dominant instabilities driving turbulence has advanced to the point where the transport produced by this turbulence can be completely suppressed. Even a decade ago, the remote control of small-scale turbulence at the 100-million-degree cores of modern plasma fusion experiments would have been unthinkable. Yet significant inroads are now being made to surmount this critical hurdle of 40 years' standing. Theoretical predictions of fusion plasma behavior have led to the design, optimization, and testing of new plasma confinement approaches and new ways of controlling the macro- and microinstabilities that limit energy containment. The close interaction between theory and experiment represents the new face of the fusion energy sciences program.

A successful scientific enterprise develops state-of-the-art research tools, uses these tools to bring rigorous closure between experiment and theory, and then innovates. The ability to diagnose experiments on high-temperature plasmas and compare the results with theoretical models and numerical simulations has improved markedly over the past two decades and in the committee's view is itself an important achievement of the field of fusion science. The enhanced ability to bring to closure some of the complex problems facing the discipline presages a new period of scientific development and supports a science-based strategy for fusion energy.

Another measure of the quality of a scientific program is the international standing of the discipline supported by the program. The U.S. fusion program has traditionally been an important source of innovation and discovery. A distinguishing feature of the program has been its goal of understanding at a fundamental level the physical processes governing observed plasma behavior. This feature, a strength of the program, was formalized in the 1996 restructuring with the new emphasis on establishing the knowledge base for fusion energy. Over the past several decades, the United States has played a dominant role in plasma theory, which is an essential tool required to unravel the complexities of plasma dynamics. The quantitative detail in which experiments are designed and executed in this country has become a benchmark for the rest of the world. The forte of the U.S. program is, as was mentioned above, the close confrontation between theory and experiment and the development of superior computational physics codes for quantitative exploration of novel physical concepts.

To assess the specific contributions of the DOE's OFES program to the fusion effort in general, one must separate the U.S. effort from the broader international effort. This is not easy, because there has been close interaction between the U.S. and international programs since the beginning, in the 1950s. To be specific, U.S. scientists played a major role internationally in developing the energy principle for describing plasma stability; heating and sustaining currents in plasmas; and understanding and controlling plasma turbulence and transport. (Specific examples of U.S. and foreign contributions can be found throughout the discussion in Chapter 2 and in Chapter 4, in the section devoted to U.S. contributions.)

In short, the quality of the science that has been deployed in pursuit of the fusion energy goal is easily on a par with other leading areas of contemporary physical science. Fusion research has mastered the ability to work flexibly with the super-high-temperature plasma state in the laboratory. It is important to note that the quality of fusion science is not universally appreciated within the broader scientific community, perhaps because fusion has been viewed as a directed energy development project rather than as a scientific enterprise. Isolation of the researchers inside the fusion program from those outside the program is another possible cause for the low opinion of fusion science despite its high quality. Most scientists funded by the program do not actively participate in the wider scientific culture. As a result, the flow of scientific information both out of and into the field has weakened. New ideas and techniques developed in allied fields are slow to percolate into the program.

All in all, a half century of research suggests that the central scientific barriers to the achievement of fusion energy will ultimately be overcome, although it is still not possible to predict when sustained fusion energy production will be realized, and much scientific and engineering work remains to be done.

PROGRAM DEVELOPMENT: PLASMA CONFINEMENT CONFIGURATIONS

Early in the fusion program a variety of containment approaches were pursued in parallel. However, the constrained fusion budgets of the past two decades were unable to support a diversity of approaches to fusion research, especially in the United States. The basic strategy during that time was to build a series of tokamak configurations of increasing size, until fusion burning became possible. The narrow program focus constrained the range of scientific questions that could be pursued but at least ensured their study in one configuration. Program decisionmaking emphasized empirical machine performance and device-specific science at the expense of results of scientific generality. Although this philosophy complicated attempts to pin down the answers to scientific questions, the ability to produce, confine, diagnose, and understand high-temperature plasmas improved significantly.

In 1985, the G7 countries sponsored an international collaboration to design the International Toroidal Experimental Reactor (ITER), reflecting optimism that continued tokamak scale-up would result in a large experimental reactor in which burning plasma science and fusion engineering issues could be addressed simultaneously. In 1996, Congress reduced OFES program funding significantly. Participation in ITER was discontinued after the ITER Engineering Design Activity was completed in 1998. The remaining international partners are working to reduce the cost of the design before deciding whether to proceed without U.S. participation.

The 1996 congressional action, a roughly $100 million cut in funding (see Appendix B), naturally stimulated reexamination of the U.S. fusion energy program. Consistent with the new science emphasis of the program, the Congress, with support from the community, repositioned the fusion program administratively and budgetarily in the same category (Function 250) as high-energy and nuclear physics and renamed it the Fusion Energy Sciences program. This shift implies that program decisions in Fusion Energy Sciences will ultimately be made using scientific criteria and standards similar to those in high-energy and nuclear physics. The committee believes that it would be harmful to the stability and morale of the program if in the near future its primary focus were to abruptly shift back to energy development.

The present study by the National Research Council is one of several that followed the 1996 congressional action. A report by the Secretary of Energy Advisory Board (SEAB)[1] was especially concerned with the relation between the DOE Office of Energy Research (OER's) magnetic fusion program and the DOE Defense Programs' (DP's) inertial confinement program; the committee has nothing to add to SEAB's conclusions in this area since it did not examine the inertial confinement program. Where the committee and SEAB comment on similar issues, they are in general agreement. Of greater pertinence here is the study by the Fusion Energy Sciences Advisory Committee (FESAC) of DOE.[2] This study, which outlines a detailed program for the next few years, has a function different from that of the present study. Nonetheless, the committee notes that the broad program goals stated in the FESAC plan are generally consistent with a science-based approach to fusion. For example, the FESAC plan recognizes that fusion research will benefit from the deployment of a variety of plasma confinement devices. On the other hand, at a more specific level, the categories of device proposed by FESAC for program planning (concept exploration, proof of principle, performance extension) continue to emphasize the evolution of specific plasma configurations toward a fusion power reactor at the expense of an understanding of the cross-cutting scientific issues. The FESAC decision criteria thus appear not to permit projects of

---

[1] Department of Energy (DOE), Secretary of Energy Advisory Board, Task Force on Fusion Energy. 1999. *Realizing the Promise of Fusion Energy: Final Report of the Task Force on Fusion Energy.* Washington, D.C.: DOE.

[2] Department of Energy (DOE), Fusion Energy Sciences Advisory Committee. 1999. *Report of the FESAC Panel on Priorities and Balance.* Washington, D.C.: DOE.

significant scale designed primarily to answer scientific questions. At this level of program detail, scientific goals are still subordinate to directed energy development goals and management conceptions.

The fusion program has undergone a surprisingly large number of reviews, with multiple levels of detail and scope, since the 1995 President's Committee on Science and Technology fusion report,[3] including the present review and the SEAB and FESAC reviews mentioned above. The multiple recommendations at various levels and the drain on program resources motivate primary recommendation 7.

In summary, the U.S. fusion program no longer concentrates solely on achieving a practical tokamak reactor in the shortest time feasible, given budgetary and other limitations, but is turning to a less time-pressured examination of the key scientific issues and technical options surrounding fusion energy. Program planning is moving away from mostly empirical decision criteria and toward scientific decision criteria. This new approach offers the prospect that rigorous fusion science will rapidly advance not only energy goals but science goals as well. However, this prospect will not be realized until identifying and answering key scientific questions become central to program planning, budget formulation, and management philosophy.

INSTITUTIONAL CONSIDERATIONS: INTERACTIONS OF THE FUSION PROGRAM WITH ALLIED AREAS OF SCIENCE AND TECHNOLOGY

There was a clear history of intellectual exchange between the fusion plasma community and the broader scientific community in the early, pioneering years of the fusion program. Much of the basic description of plasma dynamics was being developed, and many of the ideas were of sufficient generality to be widely applicable and accessible. However, in recent years the increasing specialization and technical complexity of fusion research has sharply reduced the accessibility of the work to the broader science community. This trend is evident even though many of the topics being explored could have significance for allied areas of science. The identification of the general nature of many of the more recent scientific results—and, accordingly, their potential significance to other branches of science—has not been adequately encouraged as a programmatic goal. As a consequence, the previously rich interchange between fusion researchers and other scientists has diminished greatly. The present fusion community is relatively isolated from the rest of the scientific community, seriously eroding the university base of plasma and fusion science.

A further contributor to the scientific isolation of the U.S. fusion program is its relatively narrow and static institutional base. Although the relative allocation of DOE funding among national laboratories, universities, and industries arguably is roughly the same in the fusion, nuclear, and high-energy areas, the smaller overall funding of the fusion program does not support an equally diverse community (see Appendix B for representative OFES funding data). The almost negligible funding of fusion science by other U.S. government agencies is a contributing factor, as was the focus during the 1980s on a tokamak-based, directed-energy-development program. Most important, fusion research dollars have flowed from the same agency office to a small group of universities and national and industrial laboratories that has remained relatively stable over several decades. While such long-term and sustained support can often be critical to the successful resolution of important scientific problems, it can also lead to stagnation and lack of competition. Institutional concentration and stasis have narrowed the scientific base of

---

[3] Executive Office of the President, President's Committee of Advisors on Science and Technology, Panel on the U.S. Fusion Energy R&D Program. 1995. *The U.S. Program of Fusion Energy Research and Development.* Washington, D.C.: White House.

the fusion energy science program and have acted to reduce the number of knowledgeable scientists from other fields who can contribute to fusion research.

Funding for "general plasma science" is a small fraction of the total of the fusion energy science budget, as shown in Appendix B. Some of these funds appear to have been well used; for example, the National Science Foundation (NSF) and DOE have continued to collaborate on a highly successful but small program to encourage small-scale plasma science experiments. In addition, general plasma science has found support from the National Aeronautics and Space Administration (NASA) and NSF because of its important applications to solar and space science, astrophysics, and geophysics. Industry supports major efforts in the plasma processing of semiconductor chips. Although many of the problems studied in general plasma science do not strictly pertain to the very-high-temperature fusion regime, collaboration with general plasma scientists is one of the main ways in which fusion scientists can interact with the broader scientific community. A firmer institutional commitment to general plasma science on the part of fusion energy science would build stronger links to a vibrant research community that is stimulated by a diversity of research goals.

The proportion of U.S. fusion funding devoted to competitively peer-reviewed grants is relatively small. The peer-review process is a natural way to involve a broader scientific community in the research decisions of a given field; properly administered, it can stimulate a field to evolve a healthy diversity of participants and research approaches.

The committee is concerned that U.S. fusion energy science may have a progressively narrowing demographic base. The replenishment of the fusion community in the future depends on the health of its university programs today. It is very difficult to count fusion and plasma faculty so as to estimate how many students are being trained. There is some information on physics, the discipline that gave birth to plasma and fusion research. Of a total physics faculty of roughly 1300 in 25 leading university research departments, only 3 are assistant professors in plasma physics. This small number suggests that plasma faculty in physics departments are not all being replaced. In addition, the small proportion (roughly 40 percent) of physics departments in leading research universities that have programs in plasma physics is itself a matter of intellectual concern, given plasma's status as the fourth state of matter.

2

Scientific Progress and the Development of Predictive Capability

Early experimental efforts to harness fusion energy quickly came up against the complex, nonlinear nature of plasmas. More often than not, these experiments ended with the plasma splattering against the walls of the containment vessel rather than being confined inside the magnetic bottle. Scientific advances were needed to produce a high-temperature plasma at 100 million degrees. Tools had to be developed to describe plasma equilibrium (the balance between plasma pressure and the forces of the confining magnetic fields) and plasma stability. Even after significant advances had been made on these topics, other fundamental questions remained—how large-scale instabilities cause the plasma to break up whereas small-scale instabilities cause energy leakage across the magnetic field; how an essentially collisionless plasma can be heated hot enough for fusion reactions to occur; and how phenomena at both large and small scales can be remotely measured with enough accuracy to test the understanding of plasma behavior.

A measure of the maturity of this knowledge base—and, also, of the quality of the science—is the ability to predict the performance of experiments from a fundamental understanding of the characteristic dynamics of plasmas in the laboratory and in nature. Such predictive capability goes beyond simple performance metrics such as the energy containment time, temperature, or the plasma pressure, which, although important, do not necessarily reflect the degree to which the fusion program has broadly impacted plasma science and related fields. Predictive capability also permits the confident design of future large experiments, perhaps even the optimum magnetic container (see Chapter 3 for a more complete description of the various magnetic configurations being explored within the program).

The following sections discuss progress in the understanding of plasma science and the predictive capability that follows from this understanding. The material is presented in three categories: equilibrium, stability, and transport. It must be emphasized that while the overview of science issues that follows attempts to assess some of the important accomplishments of the program as well as the challenges it faces, all areas are not given equal treatment. It is simply not possible to cover all of the scientific topics of the program in depth. In the interest of addressing at least some topics with enough detail to convey a sense of the science being done, the impact of program decisions on the science, and the scientific culture of the program, a few representative areas have been selected.

One complication in assessing the specific contributions of the U.S. Fusion Energy Sciences program to the overall fusion effort is the need to separate and compare the U.S. effort and the broader international effort. From the very beginning of the program, in the 1950s, there has been a close collaboration internationally on all aspects of magnetic confinement fusion (collaboration on inertial fusion was constrained by security issues). Large U.S. facilities have had international collaborators and vice versa. These close interactions often make it difficult to clearly separate U.S. contributions from international contributions. The science in this chapter generally refers to activities in which the United States has, at the least, played a very significant role. Where foreign programs clearly played the dominant role, this is noted.

Many of the important experimental and theoretical tools developed during the four-decade history of the program are now converging to produce a qualitative change in the program's approach to scientific discovery. Theoretical models are now sufficiently mature to describe much of the complex nonlinear dynamics of plasmas. Quantitative comparison with experimental observations is beginning to facilitate a first-principles understanding and interpretation of the behavior of plasmas. One consequence of the emerging scientific understanding of these systems is the development of techniques for manipulating turbulence and therefore controlling the energy-containment properties of the magnetic bottles. The suppression of small-scale turbulence and transport in a 100-million-degree medium is an accomplishment that is by any scientific standard a significant achievement and a sign of the high level of the science generally carried out under this program.

SUMMARY

Significant advances have been made in each of the traditional foci of plasma physics research: equilibrium, stability, heating, and transport. Over the past decade, a high level of predictive capability has been developed in several key areas. The program is moving into a new era in which the tight integration of theoretical predictions and experimental observations is enabling the control of plasma dynamics, including the suppression of turbulence and transport.

Equilibrium and Heating

The theoretical and computational tools needed for studying plasma equilibria in complex magnetic containers are now well developed and extensively used in the design of new experiments and in the analysis of existing experiments. A number of techniques, including high-power ion beams and driven waves at frequencies from kilohertz to multigigahertz, generally referred to as radio-frequency waves, have been developed to heat plasmas and also to generate and sustain plasma currents. The basic propagation and absorption physics for beams and waves are well understood. These techniques are being used to control pressure, current, and flow profiles and thus to optimize plasma performance in present-day large experiments, but techniques applicable to future high-pressure plasmas require further development.

Diagnostics for remotely measuring important equilibrium-related quantities such as plasma density, electron and ion temperatures, and magnetic field are now available in major plasma experiments. Tools to measure electric fields and associated equilibrium flows, which are increasingly recognized as having an important influence on stability and transport, are less well developed. For future novel plasma configurations, these measurement techniques will have to be extended or new approaches invented.

Stability

Pressure and current in magnetically confined plasma are limited by large-scale, ideal (zero resistivity) magnetohydrodynamic (MHD) instabilities. Standard tools for calculating these stability limits in complex magnetic geometries are widely available, and their predictions have been benchmarked with observations, aided by mature diagnostics.

Dissipative instabilities (nonzero resistivity), such as magnetic reconnection, are not as well understood because their growth rates are low. Consequently, fluid drifts and flows can strongly modify the conditions under which these instabilities begin to grow. Once reconnection sets in, however, the dynamics is better understood, and the critical role of dispersive waves acting at small scale-lengths has been identified; similar physical processes control magnetic reconnection in magnetospheric and astrophysical plasmas. Pressure-driven reconnection, which can limit the pressure in magnetic containers below the ideal limit, is inherently nonlinear and therefore challenging to model. Single-mode evolution is largely understood, but the dynamics of multiple modes, their interactions, and the related transport are not.

The existence of a density limit for plasma stability is a robust feature of magnetically confined plasmas and is well described by a simple scaling law (for tokamaks). While the final termination of discharges that exceed the density limit is described by a mechanism involving edge cooling, the apparent increase in transport at the limit is not well understood, and a predictive theory has remained elusive.

The destabilization of plasma waves by energetic alpha particles and the resulting transport of these particles and their energy are central issues for energy-producing plasmas. Experiments have verified theoretical models for stability and for the single-wave nonlinear behavior. Learning about the interaction of energetic particles with a broad spectrum of waves and the self-consistent interplay among plasma equilibrium, stability, and transport in the presence of strong local self-heating requires access to a fully ignited plasma experiment.

Transport

It has been well documented that plasma transport is mainly caused by an anomalous process in which the free energy from gradients in the density and temperature profiles drives instabilities. These instabilities lead to vortices, which in turn cause the ions and electrons to wander across the confining magnetic field and eventually escape from the container. Anomalous transport of the same sort has also been invoked in other plasma systems, ranging from Earth's magnetopause to astrophysical accretion disks.

Historically, the rate of transport and its dependence on plasma parameters were described by empirical scaling laws derived from experimental data. While the empirical laws were useful for comparing the performance of different machines and for benchmarking the testing of ideas for improving confinement, they provided little insight into physical processes. Moreover, the scaling laws had to be continually modified as plasmas progressively entered new operating regimes. This eroded confidence in the strictly empirical approach to describing transport. In place of scaling laws, a first-principles physics approach is now being developed that treats the detailed dynamics of the microscopic turbulence that drives transport. The enormous range of space- and timescales and the complexity of the magnetic geometry make the numerical simulation of turbulent transport a true "grand challenge."

Recent key discoveries in plasma transport research include the recognition that zonal flow plays a crucial role in determining nonlinear saturation and in identifying avalanches and associated fast radial propagation of disturbances. An important advance occurred when transport barriers (local regions of strongly suppressed transport) were observed to spontaneously form at the edge of a tokamak plasma.

Following a long and sustained effort to understand this phenomenon, a major breakthough occurred recently when these barriers were generated in the core of a confined plasma. The formation of these barriers marked a paradigm shift in the program, since their existence was a sign that turbulence—and hence transport—in high-temperature plasmas could be controlled. There is now substantial evidence that these barriers are generic, occurring in a variety of confinement devices. While experimental data support the idea that transport barriers arise because self-generated plasma flow suppresses the local turbulence, the actual dynamics of and threshold for barrier formation remain poorly understood. A complete understanding will require simulating barrier formation in the presence of self-consistently evolving turbulence.

Direct measurement of fluctuations is required to understand turbulence-driven transport. Experimental techniques have been developed for remotely measuring fluctuations in the temperature and density at the length of the ion gyroradius (radius of gyration around the magnetic field) or larger. However, fluctuations at the much smaller electron-scale lengths have not yet been measured. Also, the measurement of fluctuations in the electric potential and the magnetic field is difficult.

Although many instabilities can be present in modern plasma experiments, one particular instability—driven by the ion temperature gradient—has been identified as the dominant mechanism for ion thermal transport in the core of tokamak plasmas. The nonlinear behavior of this instability requires further investigation. In addition, agreement between key predictions of the models for this instability and experimental observations remains weak. The understanding of electron thermal and particle transport as well as the impact of magnetic perturbations, which are known to be important in high-pressure plasmas, also remains poor.

Fluid models have advanced our understanding of transport at the cool edge of confined plasmas. The dominant instabilities here are distinct from those in the core region. More focused experiments dedicated to comparing theory and simulation models with experimental results are needed to gain confidence in the models and to pin down the threshold for the formation of the edge transport barrier.

EQUILIBRIUM AND HEATING: DESIGNING, CONTROLLING, AND DIAGNOSING A CONFINED PLASMA

Plasmas at the extremely high temperatures and densities at which fusion occurs need to be contained in magnetic bottles that can insulate the plasma from its cold surrounding wall material. These containers must be "strong" enough to hold in the pressure of the hot plasma gas, and their magnetic field lines must have "good" topology so that hot plasma particles—especially electrons—cannot wander along the lines and escape. Typically, the field lines of good containers form closed surfaces as they wind through space, the result being that particles moving along the field lines cannot easily leave the system.

Generally, the most desirable containment configurations have large values of the parameter β, which is the ratio of the pressure of the plasma to the "pressure" (actually the energy density) of the confining magnetic field. For any given configuration, there is an upper limit on β for which an equilibrium, or balance of pressures, can exist. Above this limiting β, the plasma blows apart and strikes the walls of the confinement vessel, quenching the discharge. Hence the creation of equilibrium is the necessary first step to achieving high-temperature plasma containment.

A number of techniques, including high-power ion beams and driven waves at frequencies from kilohertz to multigigahertz, generally referred to as radio-frequency waves, have been developed to heat plasmas and also to generate and sustain plasma currents. The basic propagation and absorption physics for beams and waves are well understood. These techniques are being used in present-day large experiments for the control of pressure, current, and flow profiles in order to optimize plasma performance, but techniques applicable to future high-pressure plasmas require further development.

The theoretical and computational tools for studying equilibria in complex magnetic bottles are now well developed and extensively used. Similarly, highly developed techniques for measuring important equilibrium quantities such as the plasma density and temperature and the magnetic field are now standard on plasma experiments. Diagnostic tools to measure electric fields and associated equilibrium plasma flows are, however, less well developed.

Tools for Calculating Equilibria

The theoretical and computational tools for calculating the magnetohydrodynamic equilibria of plasmas confined in complex magnetic bottles are quite mature. The various equilibrium codes give reproducible and verifiable results for the same configuration. These codes are routinely used for the design of new experiments, analysis of data from present experiments, determination of equilibria for stability analysis, and even analysis for real-time feedback control of complex plasma shapes. Their robustness and high level of development reflect the relative simplicity of the underlying physics, which involves only Maxwell equations and the force balance for a conducting fluid in a magnetic field.

In particular, two-dimensional equilibrium codes—for configurations with an axis of symmetry—are highly developed, extensively benchmarked against experiments (especially for tokamaks, less well so for other symmetric configurations), and widely used. Some extensions are still being pursued: for example, allowing for the presence of currents outside the last closed magnetic flux surface (called the separatrix).

Three-dimensional equilibrium modeling codes are required for confinement configurations that are not axisymmetric. The outstanding example is the stellarator configuration, which has coils wound into a helical torus (see Figure 2.1). The stellarator does not have simple toroidal symmetry, although it may have helical symmetry or some useful form of quasi-symmetry. The state of the art in these codes continues to evolve, but they still provide a reasonably mature set of tools for plasma modeling and analysis. A challenge for the three-dimensional equilibrium codes is to be able to represent multiply connected regions, such as stochastic field regions (where magnetic field lines fill the volume without forming flux surfaces) or magnetic islands. Whether equilibria really exist when there are no flux surfaces is a question peculiar to the three-dimensional configurations. As long as this issue can be avoided, the equilibrium codes and vacuum magnetic field solvers provide the basic tools for design of complex three-dimensional stellarator configurations.

A modest amount of experimental benchmarking has been pursued for the stellarator equilibria, and more is anticipated in the near future. To date, equilibrium code results and two-dimensional maps of the magnetic flux surfaces and their spatial shifts are in good agreement. Much of this work has been carried out in Germany as part of the Wendelstein stellarator experiments. Not much information has yet been obtained at high β, where the pressure-driven currents are so large that the measured fields are no longer close to the vacuum fields produced by external coils. Operation at higher values of β is expected in the next generation of these experiments.

The predictive capability of equilibrium codes is contingent on the input of reliable profiles of the plasma pressure and current. The current profile can usually be satisfactorily obtained from classical calculations for current diffusion. Obtaining the pressure profile, on the other hand, requires a reliable understanding of plasma transport, which has not yet been achieved. Heuristic transport models are available, and these have often been adequate for predicting the equilibria to be realized in experiments. The equilibrium codes are also used in hindsight to interpret experimental data. In this case, the current

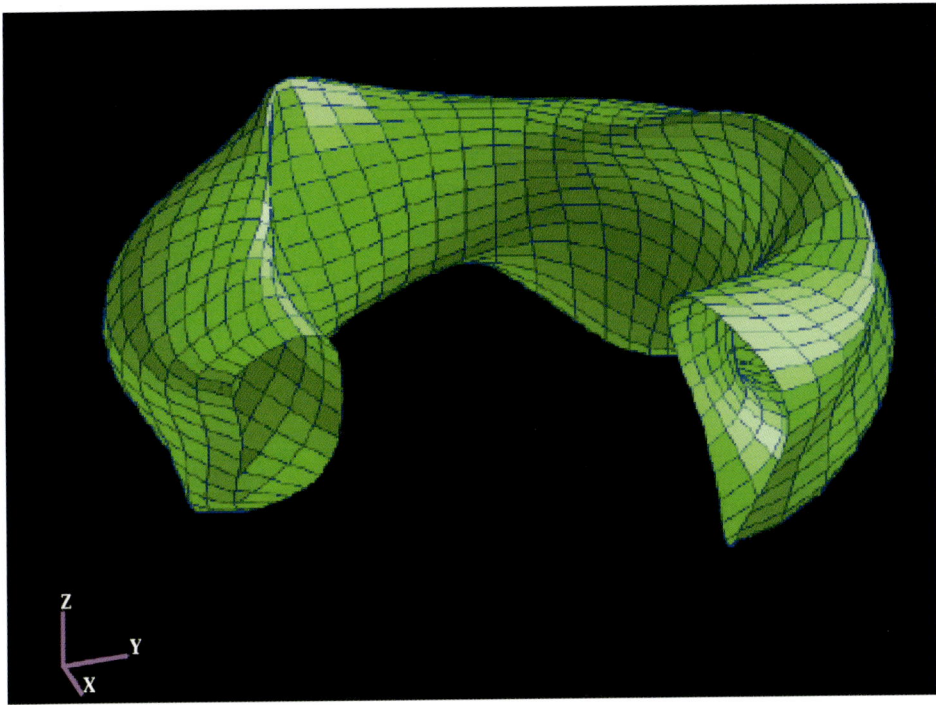

FIGURE 2.1 Pressure surface of a three-dimensional stellarator equilibrium illustrates the complex equilibria that can be studied and manipulated to maximize confinement and other properties. Courtesy of M. Zarnstorff, Princeton Plasma Physics Laboratory (PPPL).

and plasma profiles can even be extracted from measurements and applied to generate self-consistent equilibria. These solutions are then routinely utilized to analyze stability and, recently, to control the plasma with real-time feedback.

Tools for Measuring Equilibrium Pressure and Magnetic and Electric Fields

Measurements of the internal pressure and the magnetic and electric fields in high-temperature confined plasmas are critical to understanding the equilibrium and stability properties of these plasmas. Probes are useless for measuring these important quantities since they vaporize if inserted into the interior of plasmas whose electron temperature is 10 to 30 keV. Thus, the challenge has been to invent remote measurement techniques.

Very early in the fusion program, Thomson scattering techniques were developed to measure the electron temperature and density profiles. Later, when time-dependent measurements were required to understand macroscopic instability and magnetohydrodynamics (MHD), electron cyclotron emission techniques were developed to obtain accurate electron temperature measurements with high time resolution. Ion temperature measurements are more difficult and less widely available. In well-diagnosed experimental facilities, these data are provided from measurements of charge exchange between high-temperature ions and cold neutral particles.

Extensive progress has been made over the past decade in developing ways to measure the structure of the magnetic and electric fields inside magnetically confined toroidal plasmas. Most of this work was done for the tokamak configuration but is now being extended to other plasma configurations. The magnetic field at the edge of a plasma can readily be measured with simple external magnetic probes and loops. In the case of relatively cold or short-lived plasmas, the internal magnetic field has been measured with magnetic pickup loops inserted into the plasma. In general, however, remote measurements are required.

An example of successful diagnostic development is the Motional Stark effect, a remote method that can measure the internal structure of the equilibrium magnetic field, at least in plasma configurations where the magnetic field is strong. This method measures the polarization of light emitted from high-energy neutral hydrogen atoms injected into the hot plasma core region. The light is polarized as a result of the Stark splitting of the energy levels, caused by the induced electric field that the atom experiences in its Lorentz-transformed rest frame. Care must be taken to account for contributions from the intrinsic electric field in the plasma, which can often be comparable to the induced field. Such measurements typically provide spatial resolutions of a few centimeters and a time resolution of 5 to 10 ms. The tilt angles of magnetic field lines are measured to an accuracy of 0.1 deg. This accuracy is sufficient for evaluating the global stability of plasmas but not their local stability. The measurements are sufficient to confirm the existence of the theoretically predicted "bootstrap" current, which is a self-generated current present in weakly collisional plasmas. These self-generated currents aid the steady-state operation of a number of confinement approaches.

With high-temperature plasma experiments now being pursued in configurations that have weak magnetic fields and small, rapidly evolving plasmas, there is an increasing need for new or improved techniques to determine the internal magnetic field structure. In addition to the Motional Stark effect, techniques that use microwave polarimetry, emission from the ablation clouds of injected pellets, Zeeman splitting of energetic lithium neutrals, flux surface imaging, and Stark line broadening have either already been employed in special situations or are under active development. These and other concepts will have to be further developed in order to measure the magnetic field in modern high-pressure plasma experiments. New techniques are also needed to measure the profile of the current density in the cold edge region since the stability of the edge, which depends sensitively on this profile, is crucial for understanding energy containment.

Over the past several years there has been an accumulation of evidence that plasma flows, driven by electric fields that point perpendicular to the magnetic flux surfaces, can have a strong influence on plasma stability with respect to both large-scale and small-scale disturbances. These electric fields can be directly measured by a beam of heavy ions injected into the magnetized plasma. However, because it is expensive, this technique has not been deployed in large plasma experiments, where the required energy of the ions would be very high. Two alternative methods for electric field measurements in large, hot plasmas have been developed. The first measures the plasma fluid velocity and then indirectly deduces the local electric field in the radial direction from force balance. The second method uses a variation of the Motional Stark effect to simultaneously solve for the pitch angle of the local magnetic field and the local electric field strength, from which the local electric field structure can be determined with reasonable accuracy. However, the extension of the latter method to plasma configurations with weak magnetic fields is problematic and is a topic of ongoing diagnostic development.

Tools for Controlling Plasma Properties

The eventual success of any magnetic fusion approach will depend on controlling several tightly coupled processes: transport, turbulence, macroscopic stability, alpha-particle dynamics, and plasma-wall interactions. Their control, in turn, requires that one be able to control several plasma parameters: density, temperature, and the profiles for pressure, current (or magnetic geometry), flow, electric field, and so forth. Consequently, the development of techniques for plasma heating and control has always been a topic of considerable interest, including issues in plasma production, heating, current drive, flow drive, plasma stabilization, and the basic plasma physics needed for these applications. The most fully developed techniques for heating and control use either high-power neutral atomic beams or radio-frequency waves (see Figure 2.2), which are injected into the plasma. The basic physics of bulk heating and current drive has been established for the most common applications of both approaches, but techniques for detailed control of pressure and current profiles and applications to plasmas with high dielectric constants are still in development.

In neutral beam heating and current drive, energetic neutrals injected across the magnetic field deposit energy and directed momentum in the confined plasma. The generation, propagation, and deposition of

FIGURE 2.2 A gyrotron from the DIII-D tokamak experiment at General Atomics (GA) in San Diego. A gyrotron is a source of high-frequency radio waves for heating electrons. Courtesy of General Atomics.

the beam energy in the plasma are determined mainly by atomic processes and hence are well understood and predictable. Detailed measurements of the resultant ion distributions and beam deposition profiles have confirmed the models used to describe beam heating, while neutral beam current drive techniques have been reasonably confirmed with current profile measurements and detailed Monte Carlo deposition calculations. Nevertheless, the limitations of neutral beam injection for application to the large, high-temperature plasmas expected in fusion reactors have spurred work on alternative techniques for heating and control.

The interaction of radio-frequency waves with magnetized plasmas promises a wide range of plasma and fusion applications and yields a wealth of challenging scientific issues. The main issues in these applications are launching the waves, propagation into the plasma in a complex magnetic geometry, absorption of the energy (and possibly the momentum) by the plasma, and generation of the nonlinear effects that perturb the plasma and alter the wave properties.

The propagation of radio-frequency waves in magnetized plasmas is arguably one of the most quantitative and best-understood branches of plasma science. However, collisionless interactions with particles, relevant to wave heating in high-temperature plasmas, were not understood in the original studies of electromagnetic wave propagation in the 1930s and 1940s. Modern wave studies for fusion applications, begun in the 1950s and 1960s, were plagued by disappointing results owing to the poor confinement of energetic particles. Around the same time, theorists developed an understanding of wave-particle interactions (collisionless damping) and other kinetic effects. This was followed by a wealth of basic experimental work that catalogued a multitude of longitudinal plasma waves in warm, magnetized, and nearly collisionless plasmas and validated the nonlinear theory of plasma-wave interactions. As a new generation of large plasma devices with improved confinement was deployed in the 1970s and 1980s, tests of wave-plasma interactions were carried out at the megawatt power level, confirming the predictions of ion and electron bulk heating. Over the past decade, moreover, the use of radio-frequency waves to drive plasma currents at the mega-ampere level has been demonstrated at efficiencies consistent with theoretical predictions.

The applications of radio-frequency waves extend beyond bulk heating and current drive to precise and localized plasma control. For such applications to be successful, an integrated understanding must be acquired of the various aspects of wave launching, propagation, and absorption and response of the plasma distribution function to the wave. A variety of tools have been constructed for this task, especially for application to tokamak-like plasmas, but new models will be needed as new confinement schemes and plasma conditions are considered. For example, spherical torus plasmas with very high pressures have wave propagation and absorption characteristics that are quite different from those in conventional tokamak plasmas.

Launching the correct spectrum of waves into the desired plasma mode of oscillation is the first challenge. Low-frequency waves, typically associated with ion modes, require complex three-dimensional launching antennae, which are modeled numerically. The numerical models typically include some three-dimensional aspects, but a one-dimensional plasma, linear plasma response, and rectilinear antenna geometry are usually assumed. When the antenna shape does not closely match the plasma shape, the models become less accurate and less useful. An additional difficulty arises from the interaction of the antenna with the edge plasma. Nevertheless, good agreement with experimental results has been obtained for cases where the model approximations are valid. Wave-guide arrays for launching higher frequency waves, associated with electron modes, are much less sensitive to proximity to the plasma edge.

Wave propagation is modeled in one of two ways, depending on the frequency range of interest. For high-frequency, short-wavelength modes, the well-established model of geometrical optics is usually valid. At low frequencies, the wavelength is typically of the same order of magnitude as the plasma size, so the full wave equation must be solved directly. Computationally tractable two- and three-dimensional models can be formulated by means of various simplifying assumptions (such as a small gyroradius and restricting the particle cyclotron frequency to low harmonics). Such modeling yields useful and accurate predictive capability within the limits of the assumptions. However, applications to low-field devices, large reactor-scale devices, and high-pressure plasmas will require a significant increase in numerical resolution and a scale-up in computer power.

In many cases, the injected radio-frequency power does not cause the velocity distribution functions to be significantly distorted away from the Maxwellian form. Rather, it essentially provides macroscopic sources of heat, current, and/or plasma flows. Given the radio-frequency fields from the propagation models, well-developed theories can be used to calculate the local deposition of power and current. More work is needed, however, to describe plasma flows driven by radio-frequency fields. At very high powers, the distribution often deviates significantly from Maxwellian, and direct solution of the velocity-space diffusion equation is necessary to produce the resultant particle distribution functions and to calculate the associated heating and current drive. Finite difference and Monte Carlo techniques are reasonably well established for these purposes.

All of the above theories or models are computationally feasible only with idealizations and approximations, and determining or demonstrating the range of validity for a particular model is quite a challenge. To date, the most well-developed radio-frequency wave models have not been integrated into the transport and stability codes used to interpret experimental data. Tighter connection to experiments is needed.

STABILITY: PREDICTING OPERATIONAL BEHAVIOR

Fusion experiments in the early days were plagued by gross instabilities that caused catastrophic loss of plasma containment. These observations stimulated research on large-scale unstable motions. Energy principles—based initially on the MHD model of plasma as a fluid and later on kinetic models—were developed that provided a theoretical framework for understanding the plasma's stability. Subsequent experiments were then able to avoid gross instability. These energy principles are now widely used in the fusion program and also in the fields of space physics and astrophysics. Indeed, the assessment of linear stability, using computer codes that evaluate the ideal (with "ideal" indicating the limit of zero resistivity or dissipation in which the plasma and magnetic field move together, a significant constraint that maximizes the stability of the configuration) MHD energy principles in multidimensional geometry, has become a routine component of all plasma fusion experiments during both design and operation.

Modern high-temperature experiments, however, have operational limits that arise not only from ideal MHD stability but also from nonideal (nonzero resistivity) stability and from excessive radiated power (density limit). The limits associated with ideal MHD stability are typically well understood, whereas those linked to nonideal effects are still being explored. The operational limit at high density has been well characterized empirically, but a basic understanding of it remains elusive.

Also critical to predicting plasma operation, especially in configurations that could support high plasma pressure, is the ongoing effort to develop reliable codes for solving the nonlinear equations, fluid or kinetic, that describe the dynamics when linear stability boundaries are violated. In some cases, a linear instability will saturate benignly, causing only modest plasma transport that restabilizes the plasma, but in other cases, global confinement is lost. There are also purely nonlinear instabilities, which self-amplify only after the amplitude of motion exceeds a finite level and therefore cannot be analyzed with linear stability theory.

Ideal Stability of Confined Plasmas

Strong coupling between experiment, theory, and computation has, in general, led to a solid basic physics understanding and predictive capability for ideal MHD stability limits in toroidal magnetic confinement experiments. This symbiosis has resulted in innovative ideas for configurations—a notable example being the spherical torus (a doughnut-shaped axisymmetric configuration in which the major and minor radii are comparable)—that have very high pressures as measured by the plasma β. High values of β would allow fusion reactors to be compact and efficiently use the externally applied magnetic field to contain the high-temperature plasma.

The U.S. fusion program strongly supported the development of codes to evaluate local and global plasma stability from the ideal MHD energy principle. Parallel efforts were supported by the Europeans. Global stability involves unstable disturbances with large scale-lengths; these disturbances are sensitive to the profiles for the current and pressure and hence must be calculated in the full geometry of the confinement configuration. The calculation of the shape of the disturbance and how fast it grows can be reduced to a two-dimensional analysis if the configuration is toroidally symmetric. Existing codes are able to handle either "fixed" (i.e., conducting metal walls) or "free" (i.e., vacuum) boundary conditions at the edge of the plasma. The global stability codes become limited in resolution, however, when the scale-length of the perturbation becomes smaller than about 20 percent of the characteristic plasma size—in technical terms, when the toroidal mode number exceeds approximately 5. For analyzing these small-scale motions, local stability methods are faster and less computationally demanding. The best known is the so-called ballooning representation, which exploits the near-invariance resulting from the short scale-length across the magnetic field and reduces stability to a one-dimensional calculation, readily soluble even in complex magnetic configurations. The ballooning representation was developed simultaneously by scientists in the United States and the United Kingdom.

Global stability calculations based on the ideal MHD model have been quite successful in predicting the operational limits for a variety of confinement devices. For example, excessive current in a plasma can cause a kinking motion that terminates the discharge. These current limits, which were early on predicted theoretically and observed experimentally, can be avoided fairly easily by operating below the limit (in a tokamak) or by imposing a conducting shell (in a spheromak or a reversed-field pinch). Another example is the scaling for the β limit from pressure-driven instabilities, initially identified from ideal MHD calculations by European scientists in the early 1980s and soon thereafter shown experimentally in all major tokamak facilities to be consistent with the maximum achievable β value. For example, see Figure 2.3.

Present ideal stability predictions of the limits for individual tokamak discharges are accurate to better than 10 percent. These types of calculation have become essential for identifying promising innovative confinement schemes, not only in tokamaks but also in stellarators and spherical tori. Examples of discoveries that are strongly impacting the design of current experiments are "second stability" (a parameter regime in which stability—surprisingly—improves with increased plasma pressure), strong tailoring of the plasma shape, and the structure of the current profile.

Key to the successful application of ideal MHD equilibrium and stability models to experiments have been marked improvements in diagnostic and computational capabilities. Improved diagnostic measurements, especially for the tokamak, have enabled accurate reconstruction of the equilibrium current, pressure, and density profiles, which are necessary to accurately assess MHD stability. The stability of the reconstructed equilibria is now analyzed routinely and rapidly. The structure of the instabilities themselves can now be reliably obtained from measurements of the temperature, density, and magnetic fields; the disturbance in the plasma resulting from the instability typically sweeps past the

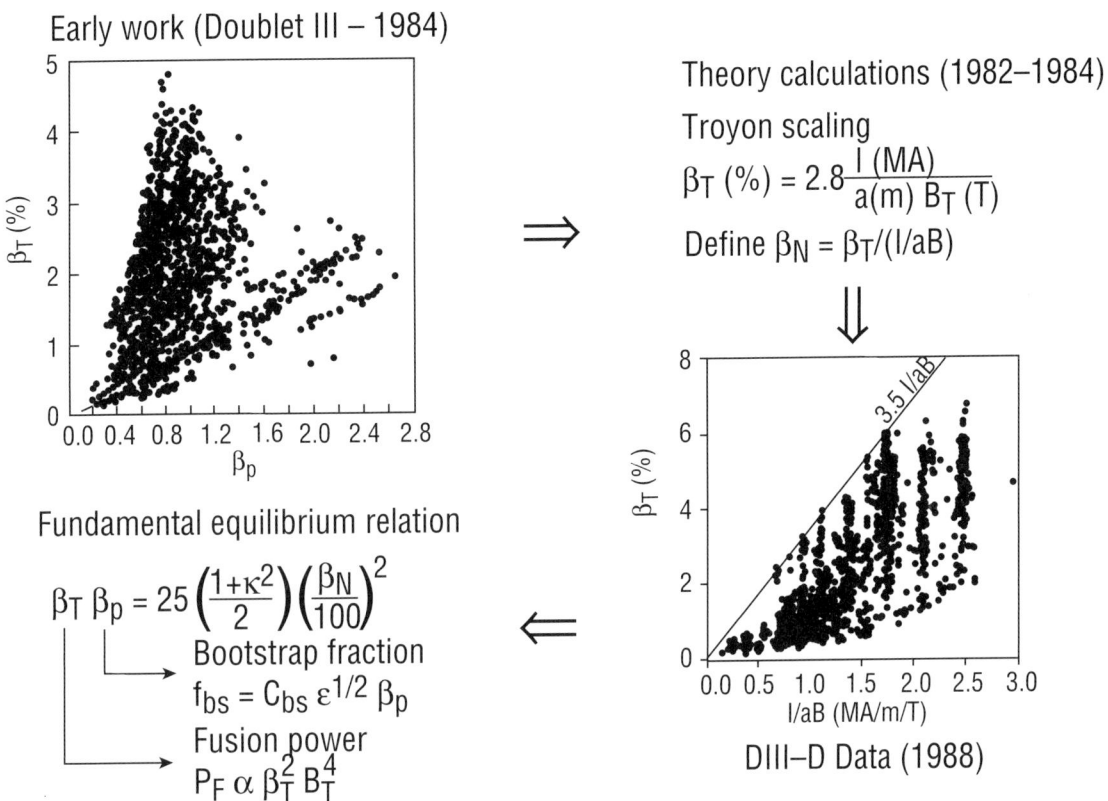

FIGURE 2.3 Pressure limits as measured by the plasma β (ratio of plasma to magnetic pressure) in the DIII-D tokamak. The convergence of the data when normalized to the theoretically based pressure limit was one of the early successes of the theory program. Courtesy of General Atomics.

laboratory frame at high velocity as a result of plasma rotation, producing relatively well-defined, time-dependent signals in many plasma parameters. Magnetic perturbations are measured with magnetic loops distributed on the outside of the plasma. The internal structure of the disturbances is obtained from soft x-ray (two-dimensional arrays are inverted to reconstruct detailed spatial structure) and electron cyclotron emission diagnostics. Improved diagnostic capabilities have spurred advances in computations, and vice versa.

Tokamak research has been the main driver for recent progress in diagnostics and in numerical predictions. Nevertheless, these tools, as well as the associated basic understanding, are generally applicable—and have been applied to some extent—to all toroidally symmetric confinement configurations. For example, ideal MHD equilibrium and stability calculations that predicted stable, very-high-pressure equilibra in the spherical torus (a U.S. contribution) have led to a worldwide campaign, first carried out in the United Kingdom, to explore this configuration. Similarly, such calculations were extensively utilized in designing a new spheromak (a spherical magnetic configuration in which the magnetic fields are largely generated by internal currents carried by the plasma) experiment. Current efforts to design a compact stellarator rely heavily on three-dimensional equilibrium and stability codes.

There are, however, three situations where ideal MHD theories are not applicable:

- *Edge-localized modes.* Tokamak plasmas operating in the H-mode (high confinement regime) have steep pressure gradients at the outer edge of the plasma, produced as a result of the development of local transport barriers. Typically these plasmas exhibit periodic instabilities, called edge-localized modes, that disrupt the barriers, leading to a loss of particle and energy confinement. Analyzing the stability of these barriers has been problematic because the current profile cannot be measured to the accuracy required to evaluate the MHD stability. Physical effects not incorporated in the ideal MHD model—such as high-speed flows, which are instrumental in the formation of the transport barriers, and kinetic effects—may also be important.
- *Resistive wall instabilities.* Almost all high-β toroidal confinement devices need a close-fitting conducting wall to ensure stability on long timescales. The wall stabilizes by blocking the penetration of magnetic flux associated with pressure- and current-driven instabilities. However, since the wall does not have perfect conductivity, magnetic flux can eventually penetrate, leading to resistive wall instabilities, which are beyond the scope of ideal MHD. In the mid-1990s, wall stabilization of a conventional H-mode plasma was experimentally shown to last up to 10 times longer than the wall resistive decay time. The theoretical explanation, that wall stabilization can be maintained if the plasma rotates relative to the wall, is supported by some experimental data. The role of the resistive wall in subduing instabilities is an ongoing subject of research. Active feedback systems are being designed to suppress the resistive wall modes.
- *Field-reversed configuration (FRC).* The FRC is a confinement device with only an azimuthal component to the magnetic field. Being compact and having inherently high pressure, the FRC is potentially interesting as a fusion reactor. Ideal MHD predicts it to be unstable. Nevertheless, experimentally the plasma remains stable over time periods much longer than the predicted instability growth time. This surprising behavior, possibly due to the large orbits of energetic plasma ions (not taken into account in ideal MHD), is still being investigated.

Nonideal Instabilities and Magnetic Reconnection

In contrast to the relatively high level of understanding and predictive capability for ideal instabilities, the understanding of nonideal instabilities remains at a primitive level. In the ideal MHD picture, the plasma is frozen into the magnetic flux, which constrains the plasma dynamics to maintain the magnetic topology. Hence, when the system starts with simple nested magnetic flux surfaces, the flux surfaces remain simple and nested. Relaxing this constraint permits magnetic reconnection, in which magnetic field lines can rip and change topology, releasing magnetic energy into energetic flows and heat and generally allowing the growth of a variety of nonideal instabilities.

Magnetic reconnection plays a major role in nearly all confinement configurations. In the FRC, magnetic reconnection is in fact required to convert the initially open magnetic system to the final state with closed field lines. In reversed-field pinches and spheromaks, the system remains in a dynamical state in which the diffusion of magnetic flux due to finite plasma resistivity is balanced by a dynamo magnetic field generation process in which the plasma evolves to a minimum energy state through a complex reconnection process. In tokamaks, periodic reconnection in the plasma core causes the central electron temperature and soft x-ray emission to exhibit characteristic "sawtooth" behavior (so called because slow buildup is followed by abrupt collapse and then repeated; see Figure 2.4). In all of these

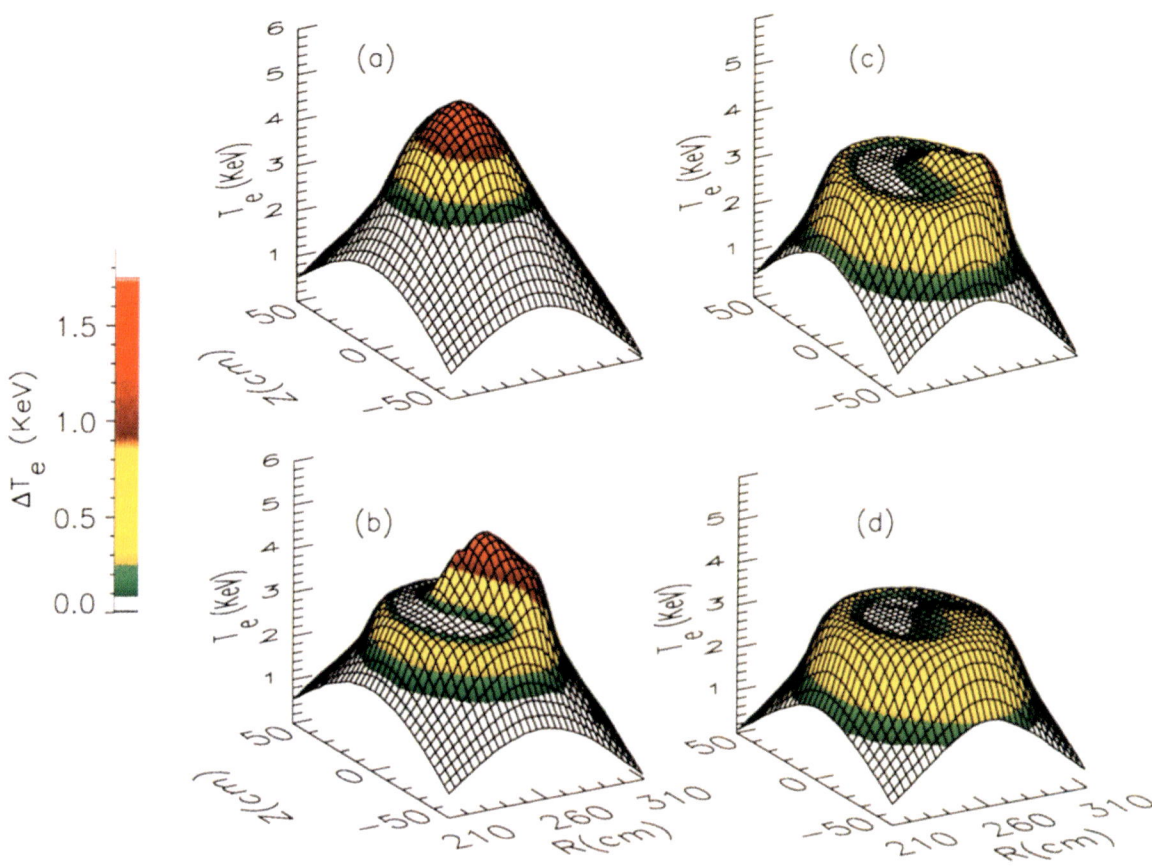

FIGURE 2.4 Measurements of the electron temperature at the core of the TFTR tokamak during a sawtooth crash. The measurements are taken with a high-resolution grating polychrometer, which measures the cyclotron emission from electrons as they spiral around the magnetic field lines. Courtesy of M. Yamada (PPPL). Reprinted by permission from *Physics of Plasmas*, 1994, vol. 1, pp. 3269-3276.

configurations, current, density, and pressure limits (termed "disruptions") are ultimately a consequence of magnetic reconnection and the associated destruction of magnetic surfaces, resulting in the loss of the plasma energy and current.

Because of its broad importance for fusion and also in astrophysical plasmas (solar flares, magnetic storms in Earth's magnetosphere, and stellar accretion disks), magnetic reconnection was widely studied by fusion scientists, especially in the early days of the program. The theory of the "tearing mode," which is the manifestation of reconnection when the amplitude of the magnetic perturbations is small, was developed for a resistive plasma. Later theories demonstrated that kinetic effects could also facilitate reconnection and that the fluid drifts of electrons and ions could significantly modify stability due to the relatively feeble growth rates of the instabilities in high-temperature plasmas. This is perhaps the primary reason why a predictive understanding of reconnection has been so difficult—the small growth rates render the instabilities sensitive to details of the equilibria, local plasma flows, and kinetic stabilizing effects.

Comparing the theoretical predictions of tearing mode growth with the experimental observations of finite-amplitude magnetic islands and disruptions from multiple magnetic islands required the development of sophisticated experimental and theoretical tools. On the experimental side, high-time-resolution measurements of soft x-ray emission and of electron and ion temperatures and densities can provide indirect data on the time evolution of the internal magnetic structure. Direct measurements of the internal magnetic field (with the Motional Stark effect diagnostic) are more difficult but have now become a standard diagnostic on major experiments. On the theoretical side, codes that solved the "reduced MHD" equations (in which the fast compressional waves are eliminated) and, later, the complete MHD equations were developed, first for simple cylindrical geometry and subsequently for the complex geometries of modern magnetic confinement systems.

The historical effort to understand sawteeth illustrates the difficulty of developing a reliable predictive capability in this area. A periodic internal instability that expels energy from the core of tokamaks was first seen on soft x-ray signals and later on other diagnostics. The explanation by Kadomtsev (a scientist from the former Soviet Union)—that current tends to diffuse into the region of high temperature and conductivity and drive magnetic reconnection, which then acts to reduce the current to a lower level—gained wide acceptance. Resistive MHD simulations reproduced the periodic sawtooth phenomena, generating confidence that theory and experiment were in agreement. That this rosy picture was not completely valid became evident with the new generation of high-temperature tokamaks, starting with the Tokamak Fusion Test Reactor (TFTR). The timescale for the expulsion of energy from the core was far shorter than could be explained by the classical Sweet-Parker magnetic reconnection theory. The mechanism for fast reconnection is now understood in terms of small-scale dispersive waves with high phase speeds that enhance rates of reconnection much above the usual MHD rates. The second surprise in the sawtooth story was the measurement of the magnetic field generated by the plasma current in the core after the energy expulsion, which indicated that the central current was not reduced during the expulsion process, in fundamental disagreement with the Kadomtsev picture. Such measurements were first completed on the Textor tokamak experiment in Germany. These observations are still not fully understood, although explanations have been offered.

Interest in sawteeth and more generally in magnetic reconnection waned after it was realized that plasma performance in tokamaks can be improved by keeping the current below the threshold for sawtooth instability. The importance of reconnection in other experiments, such as reversed-field pinches, was not sufficient to give it priority status in the fusion program. It was at that time that reconnection research shifted to the National Aeronautics and Space Administration (NASA) and the NSF-funded space science program. The recent construction of a dedicated magnetic reconnection experiment and its exploration in some of the emerging new experiments signals a rebirth of scientific interest in this area in the fusion program.

Although magnetic reconnection and resistive instabilities have historically emphasized the plasma current and its associated magnetic field as the source of free energy, there is increasing observational and theoretical evidence that the plasma pressure gradient can also drive magnetic islands and reconnection. The resulting instabilities may lead to pressure limits lower than those obtained from ideal MHD equations, which are based on zero resistivity (in this "ideal" limit, the plasma and magnetic field move together, which imposes a substantial constraint on the plasma motion). An example of such an instability, the "neoclassical" tearing mode, is associated with the currents driven by plasma pressure gradients. The neoclassical-MHD instabilities are of special scientific interest because they are nonlinear rather than linear instabilities—that is, only above a finite amplitude can they extract the free energy of the plasma pressure gradients and amplify. The growth of single modes of this sort is fairly well understood, but the nonlinear behavior and interaction of multiple modes and how they might destroy plasma containment remain poorly understood. The development of a predictive capability for these phenomena is complicated by their sensitivity to initial finite-amplitude perturbations and by the absence of a well-defined set of nonlinear equations with which their nonlinear dynamics can be explored.

An ongoing issue in the fusion program has been the degree to which the resistive MHD equations are a valid description of the dynamics that characterize modern high-temperature fusion plasmas. The two-fluid MHD description has been shown to describe magnetic reconnection much more accurately than the usual single-fluid MHD theory. In addition, the particle drifts, not contained in the MHD framework, can strongly alter stability boundaries for resistive instabilities. Insofar as the plasma dynamics are insensitive to the plasma resistivity, it may be that the resistive MHD description is adequate; however, this is not generally expected to be the case. A reasonable goal of the program should therefore be to make the transition to a multifluid model as the standard description for macroscopic phenomena. The challenge here is to be able to resolve the space scales required for accurately modeling the non-MHD waves that influence the dynamics of the system.

Density Limits

Magnetic confinement experiments operate over a finite range in plasma density, the upper end of which is typically called the density limit. This limit has been studied most extensively in the tokamak, where it is one of the fundamental operational boundaries. Indeed, for some reactor designs, performance depends critically on this limit. The density limit has also been observed in stellarator experiments, but the character, scaling, and mechanisms for density limits are much less well understood in confinement devices other than tokamaks.

In the tokamak, the limit manifests itself as a major disruption, a macroscopic MHD event that terminates the discharge. The density limit can be specified by the scaling equation $n \propto I_p/a^2$, where the density, n, is measured in inverse cubic meters, the plasma current, I_p, in mega-amperes, and the minor radius, a, in meters. While simple and based purely on empirical observations, this scaling is robust with respect to plasma size, shape, and even edge boundary conditions (how the energy outflow into the edge is treated). These variables do not strongly impact the scaling for the density limit. Significantly, the critical density does not depend on the input power or on the impurity content, as long as the latter is kept sufficiently low.

The mechanism for disruption at the density limit is not in serious dispute. As the density increases, the edge of the plasma cools, increasing the local resistivity and thus reducing the plasma current density. The current profile shrinks, making the overall MHD equilibrium unstable, and a disruption ensues. The presence of low-Z impurities in the plasma exacerbates the edge cooling at high values of the density, since they radiate more efficiently at low temperatures. This is clearly an unstable process, with lower temperatures leading to higher radiation leading to still lower temperatures in a process called radiative collapse.

While explaining the mechanism for the final collapse, this picture cannot explain the value or scaling of the density at which the collapse occurs. Density limits based solely on edge power balance and stability predict a relatively strong dependence on input power and plasma purity, as long as a sufficiently low level of impurities is maintained. This suggests that additional physics is needed to explain the density limit. Some experimental observations support this view: for example, the fast decay of plasma density following pellet injection to density values above the limit; the increase in particle transport or fluctuations as the limit is approached; and the strong dependence on collisionality of perpendicular transport at the plasma edge. From a purely operational point of view, tokamaks require a nonlinear increase in neutral fueling at high densities, an increase that tends to diverge as the density limit is approached. The connection to the power balance model is straightforward. As the limit is approached, energy and particle confinement degrade, cooling the edge and requiring ever-stronger neutral fueling for each increment in density. Ultimately power balance is lost and the edge cools uncontrollably, ending in loss of macroscopic stability.

Some theoretical support exists for this picture. Three-dimensional electromagnetic simulations of the plasma edge show a dramatic increase in turbulence at high collisionality (high density) and high β. While these simulations are at an early stage and cannot yet be used to predict the density limit, experiments near the limit do tend to operate in the regime predicted to have especially large turbulence. Calculations based on the power balance argument, when modified to include transport effects, yield results closer to those seen in experiments.

Most theories for the density limit have focused on edge power balance. As described above, these generally fail to predict important experimental observations. Although work is under way on the turbulent transport aspects of the problem, it is clear that a first-principles predictive capability is distant. Only a few turbulent transport codes show any density-limiting behavior—indeed, most turbulence simulations are run in regimes that are not applicable to the edge of high-density plasmas. Realistic magnetic geometry must also be incorporated into the codes. The advancement of theory, computer hardware, and algorithms will facilitate progress.

A predictive theory for the density limit will not emerge without strong coupling to the experimental program. This work has begun but needs to be strengthened and expanded. Capabilities for detailed profile and fluctuation measurements have been steadily building in recent years, and some machine time has been dedicated to density limit studies in several devices. The amount of experimental time should be increased to hasten progress, and the experiments will need to stress the underlying physics—substantial time has already been dedicated to empirical studies. Improved diagnostics to measure fluctuations at the plasma edge are also likely to be required.

It is probably too early to assess the transferability of theories for density limits observed on tokamaks to other types of confinement devices. However, the robustness of the limit and the simplicity of the scaling law are cause for some optimism in this regard. The lack of dependence on details of geometry, materials, and heating mechanism, as well as the unification across a wide range of machine sizes, suggests that the physics involved is common across an equally wide range of plasma regimes.

Influence of Fast Particles

The study of the role of energetic particles in plasma stability and dynamics during the past decade has been strongly focused on the development of a scientific understanding with both interpretative and predictive power. The strong focus in this area is a consequence of the significant impact of fast particles on high-performance plasmas whose temperatures approach those required for fusion. Particularly important is the open question of whether the highly energetic alpha particles that are produced by fusion reactions (e.g., the 3.5-MeV alphas in deuterium-tritium plasmas) drive instabilities, which then scatter the particles out of the discharge before they can deposit their energy into the bulk plasma. The resulting loss of energy could quench ignition (self-heating from fusion reactions exceeds the energy losses) or high-gain burn.

The most basic form of transport of energetic particles results from the drifts of particles in the magnetic geometry of the container. For very energetic particles, the dominant loss through this channel is due to magnetic field "ripple." Ripple refers to the small variations in magnetic field strength along a field line—inevitable when the field is generated by a finite number of coils—which destroy perfect toroidal symmetry. Efficient, well-validated codes that produce accurate quantitative predictions of this ripple loss are now routinely used in experimental design and analysis.

A greater challenge is the development of a predictive capability for transport due to collective instabilities. There has recently been rapid growth in the understanding of how energetic particles interact with linear modes, causing instabilities that have been observed and manipulated in experiments. The dynamics of the nonlinear system, especially when many modes interact and drive transport of the energetic particles, remains uncertain in spite of its ultimate importance in an energy-producing plasma.

A major stimulus to fast-particle studies was recently provided by the long-awaited deuterium-tritium fusion experiments on the TFTR and the Joint European Torus (JET) tokamaks in the United States and Europe, respectively. Although subignited, these experiments provided valuable information about alpha-particle behavior—for example, loss due to scattering resulting from collisions and instabilities—and plasma heating. A follow-on deuterium-tritium campaign on JET is anticipated, and studies of ultraenergetic ions from new negative-ion neutral beams on the JT-60U tokamak in Japan are already under way. Other important results have been obtained from fast-ion (and -electron) experiments in nonreacting plasma experiments that have high-power auxiliary heating. However, it may be argued that without a true burning plasma facility, the experimental tools for fast-particle physics studies are inadequate.

The availability of new fast-particle diagnostics enhanced the productivity of the deuterium-tritium and other fast-ion experiments and also facilitated the comparison of experimental data with the predictions of theoretical analysis and computational studies. Specific examples are fast-ion loss detectors, charge exchange spectroscopy, and pellet charge exchange. Although initially developed for the deuterium-tritium experiments, these have now been installed on facilities elsewhere.

Researchers have developed single-particle, kinetic, and hybrid fluid-kinetic theoretical models for fast particles, which have on the whole been adept in describing their behavior. Hand in hand with fast-particle theory there has been progress in developing and applying sophisticated numerical codes. Indeed, the synergism of analysis, computation, and experiment (along with the programmatic focus on high-performance plasmas equipped with modern diagnostics) has been responsible for the early successes achieved in this field.

Particularly successful has been the interpretation of collective instabilities, ranging in frequency from 10 Hz to 10^9 Hz, which can be excited by fast particles. The experimental observations of the kinking of the core of tokamak plasmas, called "fishbones" because of their characteristic signal on the magnetic pickup coil diagnostics, were explained as being due to destabilization by fast-beam ions, which resonate with the disturbance. On the other hand, off-resonant fast ions were predicted to stabilize the kinking, and this was also observed in experiments. The fast-particle excitation of Alfvén waves has also been identified in dedicated experiments and successfully simulated numerically. In general, the frequencies and mode structure obtained theoretically match the experimental measurements quite well, while the stability thresholds, more difficult to calculate (owing to sensitivity to pressure and current profiles), are in reasonable agreement.

The nonlinear saturation levels of these collective instabilities and how they affect alpha-particle transport are much more difficult to calculate. Initial loss estimates of alpha particles resonant with Alfvén waves were obtained from test-particle simulations using an externally imposed mode structure. Self-consistent theories that allowed the wave amplitudes to evolve showed subsequently that saturation occurs due to the redistribution of resonant particles by finite amplitude waves. Particularly interesting when the mode is barely above the threshold is that its amplitude can grow explosively and persist for several hundred inverse growth times, while its frequency bifurcates and shifts both upward and downward. Theory and simulations have shown that such behavior is a consequence of the trapping of the energetic ions in the wave field. This could provide a unified explanation for the strong frequency chirping also observed in other plasma devices and even in high-energy particle accelerators. Already the nonlinear frequency splitting (called "pitchforking") observed in the JET tokamak has been interpreted by this theory.

Whether the present understanding of fast-particle physics can be extrapolated to large burning plasmas whose behavior is dominated by alpha-particle self-heating is questionable. In this new regime, stability and confinement theory will need to be generalized. One critical issue for thermonuclear fusion devices concerns very small-scale, unstable Alfvén modes, which could be a potential loss mechanism for alpha-particle energy. Codes that specifically address the linear stability of such modes are being developed, although they are currently limited by being computation-intensive. Extending nonlinear codes to the reactor-relevant regime of small gyroradius (relative to plasma size) will require going beyond single-mode analysis to describe a "sea" of resonantly overlapping short-wavelength modes. Present-day experiments cannot destabilize a sufficient number of modes to explore this regime.

TRANSPORT: ENSURING SUFFICIENT CONFINEMENT

Once models for the equilibrium and large-scale stability of magnetically confined plasmas had been established, theorists began exploring how classical interparticle collisions cause energy and particles to be transported across the confining magnetic field. The incorporation of magnetic drift motion in the particle orbits led to a generalized description called neoclassical transport theory. This theory describes the irreducible minimum rate of collisional transport of particles and energy across the confining magnetic field in toroidal geometry. The essential ingredients of this theory were identified by scientists in the former Soviet Union. It soon became clear, however, that the leakage of energy out of magnetic bottles greatly exceeds the predictions of this theory. The larger-than-expected value of the transport rate was named "anomalous transport." The culprit for anomalous transport was presumed to be short-wavelength instabilities driven by gradients in the plasma temperature and density. These gradients provide sources of free energy to generate electric and magnetic field fluctuations, which anomalously enhance transport across the magnetic field.

Understanding the phenomena of anomalous transport has been one of the grand challenges of plasma science. On the theoretical side, this effort is complicated by the difficulty of understanding, in complex magnetic geometries, the linear growth and nonlinear saturation of instabilities that evolve into a fully turbulent state; by the tremendous range of space- and timescales associated with this turbulence; by the extreme anisotropy of the fluctuations parallel and perpendicular to the confining magnetic field; and by the essentially collisionless character of the plasma dynamics in the core of most confinement systems. On the experimental side, the difficult issues are how to remotely measure the spectrum of small-scale fluctuations in the density, temperature, and electric and magnetic fields; how to determine the linear growth and nonlinear mode coupling properties of a fully saturated turbulent plasma; and how to visualize this turbulence (as has been so effectively done in experimental studies of fluid turbulence).

In the early exploration of the anomalous transport problem, the failure of simple models based on the identification of the characteristic space- and timescales of the small-scale fluctuations led to the realization that it would be immensely difficult to develop first-principles models. There was strong sentiment that even with substantial support, the problem was simply too complex to be soluble, at least with the array of computational and experimental tools available at the time. There was also a strong belief that the development of empirical scaling laws, deduced from how the energy confinement varied with plasma and configuration parameters in existing experiments, would be sufficiently accurate to predict the performance of future machines. The analogy with the development of commercial aircraft was often invoked—namely, that a first-principles understanding of three-dimensional fluid turbulence had not been required to design and build efficient aircraft. The dramatic advances in the ability of scientists to model turbulence in complex magnetic geometries, combined with the failure of the empirical approach to account for phenomena such as the spontaneous formation of transport barriers, has caused the empirical approach to come under fire and has provided increased confidence that a true predictive capability may be achievable.

Another driver for the development of confinement predictability is the sensitivity of performance projections to uncertainties in scaling laws. In scaling up to burning conditions, which are sensitive to core temperature and density, the inability to form an edge transport barrier would be an unacceptable setback. A firm first-principles understanding of all the factors controlling the formation of barriers is therefore essential.

Empirical Scaling Law Approach

Historically, studies of the confinement properties of plasmas in tokamaks and other magnetic configurations have been based on scaling laws deduced from the measured variation of the energy confinement time with plasma parameters. This procedure was grounded in necessity, since a first-principles theory of the transport of energy across a magnetic field due to plasma turbulence had not yet been developed. It also mirrored approaches used in fluid dynamics and engineering fields where basic theory was either lacking or not fully developed.

Two fundamental assumptions defined the pervading scientific thought on energy confinement at that time. First was the idea that, at some level, anomalous transport is intrinsically determined by natural processes beyond the control of scientists. The second was that a set of scaling laws could be deduced that would be universal for a given plasma configuration. Both assumptions would eventually be proven false as the richness and complexity of the transport problem became more evident and experimental techniques evolved.

The first of the empirical scaling laws, called Alcator scaling, resulted from a set of experiments on the ohmically heated Alcator A tokamak in the late 1970s. The energy confinement time in this device was found to be proportional to the product of the plasma density and the square of the small radius of the doughnut-shaped plasma configuration. Confirmation that other ohmically heated tokamak experiments satisfied the same energy confinement scaling law reinforced the view that this law was in some sense universal for all tokamaks.

A big surprise came with the construction of the Alcator C machine, which was basically a larger version of the earlier device. In the new experiment, the energy confinement was lower than had been predicted by the Alcator scaling, especially at high density. The culprit was the density profile, which in the new machine was less peaked in the central region as a result of the shorter mean free path of the neutral source at the edge of the plasma. Injecting frozen pellets to "fuel" the core region restored good confinement. Later experiments in the TFTR tokamak reinforced the conclusion that the density profile can impact confinement. Only when the influx of neutral particles from the edge is controlled, leading to very peaked profiles, can good confinement be achieved. These were the first experiments to suggest that instabilities driven by the ion temperature gradient are a determining factor in core energy confinement.

The prevailing view of anomalous transport underwent a fundamental paradigm shift with the discovery of the high confinement (H-mode) regime of operation in tokamaks in the early 1980s. As was first discovered in the ASDEX tokamak experiment in Germany, with increased energy input (by ohmic heating, neutral-beam heating, or radio-frequency wave heating) the plasma discharge spontaneously enters a new regime in which the pressure profile becomes very steep near the edge, raising the stored energy over the entire central plasma and approximately doubling the energy confinement. Nearly a decade passed before the essential ingredients of the H-mode transition were understood—namely, that plasma rotation, locally generated at the edge, shears apart the vortices driving transport, allowing the pressure gradient to steepen, which facilitates a further increase in the plasma rotation. Nevertheless, once the results were confirmed by other experiments, there was immediate recognition of the far-reaching conclusion that anomalous transport is not intrinsic to a magnetically confined plasma but can be manipulated if the appropriate control knobs are identified.

The discovery of the H-mode further eroded confidence in the scaling law approach, since no scaling law could predict bifurcation, in which the plasma confinement makes a transition from one regime to another. Other discoveries with significant implications for confinement were also made: injecting pellets or altering the twist of the central magnetic field (magnetic field shear) can induce a transport barrier in the core of the plasma; supplying auxiliary power on top of the ohmic heating degrades confinement (this unfavorable scaling of confinement with energy input is referred to as L-mode scaling); injecting trace amounts of heavy ions can boost confinement; the shape of the plasma cross section can affect confinement; and toroidal rotation induced by the injection of high-energy neutral beam ions can tear apart turbulence, as in the H-mode transition. The scaling law approach was improved somewhat by using normalized (or dimensionless) local plasma parameters as the variables against which the plasma confinement is compared. Nevertheless, scaling laws for describing energy containment carried more and more qualifiers, requiring that only discharges with, say, matching shapes or density profiles could be compared.

The growing complexity of anomalous transport from the experimental perspective, coupled with the opportunities provided by an array of control techniques, undermined the scaling law approach to confinement predictability. At the same time, the ability of scientists to measure, understand, and model plasma turbulence was evolving rapidly, providing hope that a truly physics-based approach to understanding anomalous transport could be successfully pursued.

Development of Tools for Calculating Stability and Simulating Nonlinear Microturbulence

In the early years of fusion science research, theories for anomalous transport used simple mixing length estimates based on linear instability properties. These estimates assume that during a few growth times, the turbulence increases to a level sufficient to mix or relax the gradients (of, for example, density or temperature) that drive the turbulence. This approach was generally inadequate to describe the observations or provide useful predictions—estimates of the linear growth rates of instabilities were often inaccurate and the postulated fluctuation levels did not reflect the nonlinear processes that are important in high-temperature plasma systems. A variety of theoretical tools therefore needed to be developed to have a useful description of small-scale turbulence in high-temperature plasmas—tools to understand the linear stability of fluctuations and also to understand the nonlinear behavior.

The unstable fluctuations of interest are those excited by the gradients of density, temperature, and velocity flow and also by the presence of multiple ion species, as often is the case in current experiments and will be in future experiments. Analyzing linear stability, especially in a complex magnetic geometry, is nontrivial. The ballooning mode formulation, which exploits the local invariance properties of plasmas with respect to translations across the magnetic flux surfaces, allows a two-dimensional linear stability problem to be reduced to a one-dimensional system. The resulting equations can be solved even in a kinetic representation. Codes based on the ballooning mode formulation, which have been benchmarked to ensure reliability, are now widely available. They are flexible so that stability can be explored not only for model plasma equilibria but also for the equilibria obtained from actual experiments. Stability results can therefore be used either to gain a fundamental theoretical understanding of a problem of interest or to analyze the stability of particular regions of the experimental profiles to help interpret the observations.

Describing the nonlinear, three-dimensional turbulent state is even more challenging. Nonlinear simulations are required, the computational algorithms for which are still under development. A wide range of instabilities can exist in a magnetically confined plasma, and their interaction is usually complicated and not well understood. A particular challenge has been the treatment of the relatively fast motion of electrons parallel to the magnetic field in conjunction with the much more slowly moving ions and the relatively slow evolution of the turbulent spectra. Thus, taking into account the perturbations of the magnetic field, which depend sensitively on parallel electron currents, is computationally prohibitive in the high-temperature core of a plasma. At the cold plasma edge, codes based on less demanding fluid-type models can be used. The description of the hot, collisionless plasma core region has therefore been largely limited to electrostatic turbulence, in which magnetic field perturbations are neglected. A few exceptions to these generalizations are discussed below.

Two key advances were important for developing the capability to simulate turbulence in magnetically confined plasmas. The first was the formulation of a reduced set of equations for which the motion of plasma particles as they gyrate around a magnetic field line, which is very rapid compared to all frequencies characterizing turbulence associated with pressure-gradient-driven instabilities, is averaged away. These so-called gyrokinetic equations can be solved without the time step constraints (and the computer memory storage requirements) imposed by the fast gyromotion. A further reduction led to a set of "gyrofluid" equations, which, although fluid in nature (that is, they are based on velocity moments that integrate out the wave-particle resonance information), still capture some of the collisionless dissipation contained in the full kinetic equations. The advantage of the gyrofluid equations is that they can be solved much more quickly than the gyrokinetic equations. Recently, however, the utility of these gyrofluid equations has been called into question because they do not adequately treat the damping of plasma rotation. Nevertheless, these equations did play an instrumental role in developing the first credible models of the turbulence driven by the ion temperature gradient instability in tokamak plasmas.

The second key advance in the ability to simulate fine-scale turbulence was the development of "flux tube" codes in the early 1990s. These codes took advantage of the same near-invariance properties as in the ballooning mode representation in order to be able to describe accurately both large scale-lengths (1 m) along the magnetic field and short scale-lengths (0.001 m) across it. For the first time, turbulence in three-dimensional systems could be simulated with normalized parameters (for example, the ratio of the ion gyroradius to the plasma size) that correspond to realistic experimental values. These flux tube codes led to the first simulation-based predictions of the transport of ion energy in the core region of magnetically confined plasmas.

While electrostatic turbulence has been the focus of study in the high-temperature interior of a plasma, magnetic field perturbations have been explored in two contexts: at the plasma edge and in very-short-scale turbulence driven by the electron temperature gradient. At the low-temperature plasma edge, since collisions are important, three-dimensional models of turbulence based on the collisional fluid equations were derived. These models have also been used to study the pedestal in the temperature profile at the plasma edge, which occurs during the transition to H-mode behavior. Because they are simpler than kinetic models, these edge-turbulence models can describe magnetic perturbations, as well as include both ion and electron temperature gradient drives for instability and correctly treat the neoclassical damping of flows, which is important for studying how stabilizing flows are generated during the H-mode transition. In a sense, therefore, these edge-turbulence models are the most complete of all plasma turbulence models.

Magnetic field perturbations and the required dynamics of the very-high-speed electrons parallel to the magnetic field have also been included in studies of short-wavelength turbulence driven by the electron temperature gradient in the hot plasma core. Since the characteristic scale-length of this turbulence ranges between the electron gyroradius and the electron electromagnetic skin depth and is therefore very small compared to the ion gyroradius, static behavior for the ions is a valid approximation. With ion-scale turbulence neglected, it becomes computationally feasible to explore the turbulence spectrum and the associated plasma transport.

Development of Tools for Remote Measurement of Fluctuations and Transient Phenomena

On the experimental front, the evolving understanding of plasma confinement has been paced by the development of sophisticated diagnostics that can remotely measure hot plasma properties. For instance, the understanding of the dominant form of transport, that of ions, as well as the ability to suppress that transport, became possible only after new techniques were invented (mainly in the 1980s) to measure in detail the ion temperature, plasma flow velocity, and plasma current profile. Similarly, to fully understand microturbulence and its relation to the anomalous transport of heat and particles requires remote measurements of local fluctuations in density, temperature, magnetic field, and electrostatic potential. Impressive strides have been made, but further development of diagnostic tools is needed to allow detailed comparisons with turbulence theory.

Early diagnostics of plasma turbulence used electromagnetic wave scattering to measure density fluctuation amplitudes over a wide range of wavelengths. Though limited in spatial resolution, these measurements did show that the power increased with the fluctuation wavelength. There was, as a result, a concentrated effort to develop new techniques to measure fluctuation spectra in the region of long wavelengths (i.e., larger than the ion gyroradius). Techniques based on the reflection of microwaves at critical density layers and on the collisionally induced fluorescence of injected neutral atomic beams now provide detailed measurements of density turbulence in high-temperature plasmas. The amplitudes of turbulent fluctuations can be measured to within 0.1 percent (relative to the equilibrium density value), and their temporal and spatial correlations are available for wave numbers from 0.1 to 10 cm^{-1}.

Temperature and also electric and magnetic field fluctuations are less well diagnosed than density turbulence. Under some circumstances, local fluctuations in both the ion and electron temperatures have been measured, allowing theoretical predictions to be tested. Even more difficult are measurements of the fluctuating electrostatic potential and magnetic field; these have been done only in particular experimental devices. Electrostatic potential fluctuations have been measured in small tokamaks and stellarators by means of the heavy-ion beam probe. Magnetic fluctuations have been measured in the reversed-field pinch device, where they are especially large (~1 percent of the equilibrium magnetic field).

Though incomplete, these new measurement techniques have already provided a wealth of information about plasma microturbulence. The peak in the wave number spectrum for density turbulence was identified and the measured spectra were fairly well reproduced from gyrokinetic calculations. A specialized diagnostic was able to demonstrate that ion thermal fluctuations have larger amplitudes than ion density fluctuations, confirming a crucial theoretical prediction related to the ion-temperature-gradient instability. Turbulence correlation lengths and times estimated from simple random-walk transport models with step size given by the turbulent eddy dimension and time by the eddy correlation time can account for the observed cross-field transport. These findings have led to the conclusion that ion transport loss is probably caused by drift wave turbulence from ion-temperature-gradient instabilities. Electron turbulence is not as well diagnosed; only recently has this transport channel received new attention.

Detailed comparisons with theory will require further advances in experimental technique: for example, in nonlinear spectral analysis (to measure growth rates); two-dimensional visualization of turbulent density fluctuations; high-resolution measurements of intermittency; and very high spatially and temporally resolved studies of transport barriers and how they suppress turbulence. Key ingredients for the formation of transport barriers are self-generated plasma flows and fluctuations in the electrostatic potential, whose exploration will require new methods to measure flow velocity fluctuations. In order to visualize turbulence, probe arrays are being developed for cold plasmas and the edge of hot plasmas, while several techniques—neutral beam fluorescence imaging, multidimensional microwave reflectometry, and phase contrast imaging—show promise for the high-temperature interior region. Nonlinear spectral analysis may soon be able to measure the transfer of energy among fluctuating modes, which will make it easier to confront theory with experimental results.

Transport Barriers and Confinement Control

The discovery of the H-mode transition—and the associated sudden improvement in global plasma confinement—marked a paradigm shift in the fusion science program. It was proof that anomalous transport is not intrinsic to a magnetic confinement configuration but can actually be reduced. First discovered on the ASDEX tokamak in Germany, the H-mode was soon thereafter observed experimentally in the United States and around the world.

More daunting was the task of identifying the cause of the H-mode transition, in which transport at the plasma edge bifurcates, leading to the creation of a narrow layer ("transport barrier") within which the density and temperature profiles are very steep. New diagnostics, which were needed to measure the fine structure of the barrier and the local gradient-driven turbulence, revealed that the fluctuations driving transport drop precipitously at the location of the barrier. This observation led to sheared plasma rotation being identified as the mechanism for suppressing turbulence. Strong support for this conclusion was provided by experimental measurements indicating that these flows develop immediately before the barrier forms. Finally, the bifurcation nature of the H-mode transition is explained by a feedback loop in which suppression of turbulence leads to local steepening of the pressure profile and enhancement of the sheared rotation. A complete three-dimensional computer simulation of the bifurcation (based on a fluid model), including the turbulence and self-generated flows, further confirmed this picture, as illustrated in Figure 2.5.

Another paradigm shift occurred with the discovery that a transport barrier can also be formed in the hot plasma core. Worldwide, experimentalists found that by manipulating the twist of the magnetic field and injecting frozen hydrogen pellets they could form "internal transport barriers" and suppress turbulence. Conversely, in experiments where the plasma rotation was created by injected neutral beams, a reduction of sheared rotation (by adjustments to the beams) caused the internal barriers to disappear. The result from experiments on the DIII-D tokamak—that anomalous ion transport was completely suppressed over the entire cross section of the plasma—cemented the current view that turbulence and transport can in fact be controlled. The theoretical picture of transport barriers now needs to be tested in sufficient detail to achieve predictive capability. Additional challenges are to extend ion transport suppression to the electron loss channel and to ensure that the resulting pressure profiles are MHD-stable and sustainable in steady state (i.e., not transient).

Transport barriers are not unique to the tokamak configuration but appear to be a generic phenomenon in nearly all magnetic confinement systems. The stellarator configuration, in which external helical coils

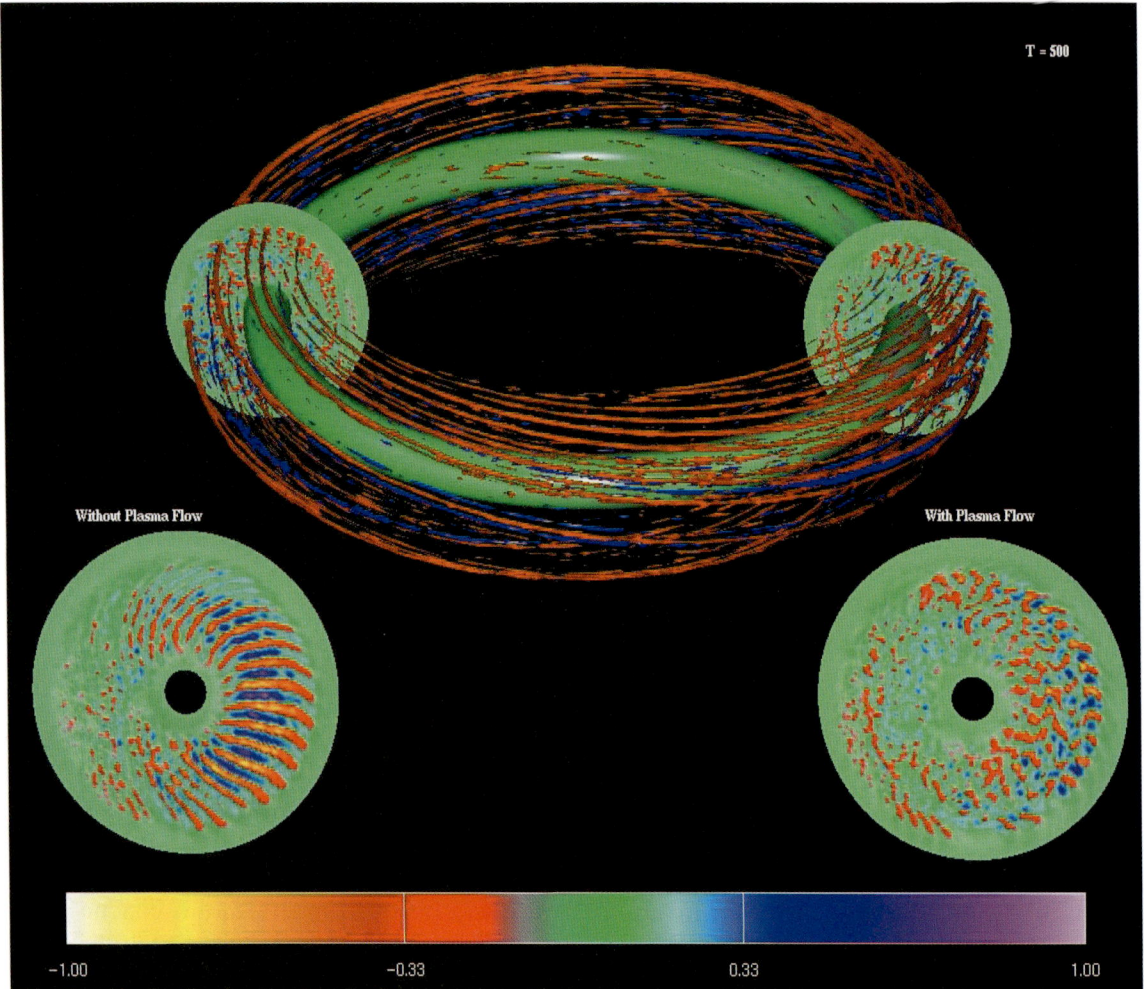

FIGURE 2.5 Simulation of ion energy transport in a tokamak plasma. Surfaces show the vortices that develop both with and without large-scale poloidal rotation, which acts to break up the flow patterns, reducing transport in magnetic confinement systems. Courtesy of W.M. Tang (PPPL). A similar image first appeared in *Science*, 1998, vol. 281, pp. 1835-1837.

(rather than internal plasma currents, as in a tokamak) are responsible for creating the equilibrium magnetic field, is currently a major focus of the European and Japanese fusion programs owing to its intrinsically good stability and steady-state nature. In both European and Japanese stellarator experiments, the H-mode transition has been observed. An important component of the reversed-field pinch experimental program is the manipulation of transport by driving currents to maintain the configuration in its minimum energy state. Improvements in confinement by a factor of 6 were documented. There is also evidence that a transport barrier induced by sheared flow can form at the edge of a reversed-field pinch plasma.

Although the basic idea that sheared rotation can induce transport barriers and reduce turbulence is now well established, this understanding has not yet been translated into full predictive capability for the formation of these barriers or for their properties. Balancing the calculated linear growth rate of relevant instabilities with the estimated damping rate induced by the measured sheared rotation has had some success in predicting when internal barriers form. However, in notable cases, this procedure fails.

Direct numerical simulations of the internal transport barriers with the turbulence evolving in time have yet to be completed. Such simulations are necessary to understand the full dynamics of barrier formation. They may also shed light on how barriers affect electron transport—experimentally, electron energy confinement sometimes improves when a barrier forms, but sometimes it does not. Simulations of edge barrier formation during the H-mode transition have been completed, and they give onset criteria that show some correspondence with the observations. Still, the sensitivity of the H-mode transition to factors such as the direction of the toroidal magnetic field has not been convincingly explained, which calls into question the current understanding of barrier onset.

Evaluation of the Present Understanding of Turbulent Transport

The understanding of turbulent transport has evolved rapidly over the past decade, driven by the development of computational techniques for directly simulating turbulence and the increasing precision by which turbulence and transport in experiments can be accurately measured and manipulated. The greatest progress has been made in the understanding of ion thermal transport in the plasma core and transport in the colder edge plasma of tokamaks. Core particle transport and electron energy transport remain poorly understood. Although confinement in tokamaks has received the most attention, the tools developed to study this problem are general to any toroidal magnetic confinement system. With the growth of interest in nontokamak approaches, it is expected that transport in these other configurations will be explored more thoroughly in the near future. Already, for example, efforts have been made to address the confinement issue in the reversed field pinch.

A number of important, fundamental discoveries have been made in the effort to understand plasma turbulence and transport. These discoveries include the identification of the short-wavelength instabilities driven by ion temperature gradients as the most probable mechanism for anomalous ion transport; the self-generation of zonal flows as the primary mechanism for saturating the linear growth of instabilities driven by magnetic field line curvature; the creation of avalanches, associated with fast radial propagation of heat pulses and cold pulses; the role of velocity shear fields in stabilizing local turbulence; and the self-generation of transport barriers, which locally suppress turbulence and transport. It is important to call attention to these discoveries because they could have broad importance beyond the fusion science program. At the same time it is important to note that some of these concepts have roots in other areas of physics and are not wholly the invention of the plasma community.

There is now substantial, but not overwhelming, evidence that turbulence driven by the ion temperature gradient is the dominant mechanism for ion thermal transport in the core of tokamak plasmas. It has long been recognized that, above a certain threshold, the ion temperature gradient could drive turbulence of relatively long wavelength (10 ion gyroradii) and that such large-scale turbulence would drive substantial transport. The observed sensitivity of transport to the density profile in the Alcator C and TFTR experiments was qualitatively consistent with a model of transport based on this instability. Measured wave number spectra are consistent with simulations of this instability, although much more detailed comparisons of theory with experimental results should be completed. An anomalous transport rate predicted from three-dimensional turbulence simulations gave a global energy confinement time that agreed well with experimental data. This comparison provided the first evidence that direct numerical simulations could compete with empirical models in reproducing the confinement trends in present experiments. Finally, the ion-temperature-gradient instability has the property that it produces stiff profiles—that is, because the transport rapidly becomes very large when the temperature gradient exceeds a threshold, typically the gradient remains near the threshold. The stiffness of the ion temperature profiles has been confirmed in experiments that demonstrated a direct correlation between the central and edge ion temperatures.

In spite of the substantial effort to test and document the ion-temperature-gradient picture of core confinement, the weight of confirming experimental evidence remains surprisingly and unacceptably thin by normal scientific standards. The verification of a transport theory on the basis of its steady-state predictions is too far removed from the basic physics to be of great value. An ongoing effort to compare the predictions of theory with an oscillating energy source in the DIII-D tokamak experiment is a positive development, but it is still somewhat indirect. Much more direct comparison between predicted turbulence properties and detailed measurements is needed to confirm the theoretical picture being developed. To date, measurements have concentrated on first-order spectral properties of the local turbulence, such as fluctuation amplitudes, correlation lengths and times, and corresponding wave number spectra, usually in one spatial dimension. Consistent data sets and relevant parametric scans of even these first-order quantities are very rare. For example, turbulence properties as a function of the critical ion-temperature-gradient scale-length are not available from any experiment. There is a need for expansion of diagnostics to measure higher order spectral properties that can provide direct quantitative tests of local growth rates and stabilizing shear flow rates and direct measurements of energy flow between fluctuating modes to evaluate nonlinear energy transfer between stable and unstable modes. In addition, multifield (e.g., density and electrostatic potential, or temperature) turbulence measurements are required to directly measure transport.

An ongoing question is how the very large volume of data now available from the simulation community in the form of movies and other diagnostics can be compared with measurements from real experiments. Visualization is a powerful tool for comparing theory and data, at least qualitatively. Data from simulations very clearly reveal the development of extended radial fingers associated with the curvature-driven instability. Self-generated plasma flows (i.e., zonal flows) break up the radial streams. Suppression of these zonal flows in numerical simulations causes the transport to rise by an order of magnitude; accordingly, the importance of zonal flows in controlling the fluctuation levels is widely accepted. Experimental efforts in visualization can help confirm or challenge theoretical predictions. Initial results from visualization experiments are just now becoming available on a limited scale. In pursuit of these goals, there needs to be renewed commitment to the development of appropriate experimental tools and to dedicated experimental run time.

On the theoretical side, the key ingredients that control the saturation of the ion-temperature-gradient instability and its wave spectrum are not fully understood. There are no theoretical predictions of the shape of the wave spectrum as the instability exceeds the threshold. At the threshold the spectrum is peaked near the ion gyroradius scale, and above the threshold the spectrum shifts toward long wavelengths—but how this shift evolves has not been quantified. Does a cascade process drive it, or is the long-wavelength turbulence locally driven? Does the measurement of the fast propagation of hot and cold pulses imply that studies of local transport are invalid? Is transport dominated by avalanches? These basic questions remain to be answered. Sharp scientific questions about the nature of the turbulence must be formulated and their answers pursued. An emphasis on simply producing a transport rate is superficial and is inhibiting the attack on these central issues and reducing the scientific impact on other fields. The impressive powerful computation tools that have been developed in the fusion science program to model turbulence should be fully used to address more of the fundamental processes that control plasma turbulence. Addressing such issues is necessary to achieve predictive capability.

Evidence is accumulating that the energy loss through electron scale-length modes is substantial and cannot be neglected in comparison with that through the ion channel, especially since control techniques for the ion channel have been developed. In some cases, the electron energy transport exceeds that through the ion channel. For example, an analysis of core transport barriers reveals that even when the ion energy transport is suppressed, the electron energy transport is not and may remain large. Surprisingly, electron thermal transport in the core of the TFTR tokamak remained large even though the electron temperature gradient was essentially zero and the density gradient was nonzero, a configuration that is apparently stable to all linear modes. Can a nonlinear instability sustain turbulence in the core of tokamak plasmas? This is a challenging but important research topic.

As discussed earlier, the inclusion of electromagnetic perturbations and the evaluation of electron thermal and particle transport are difficult because the full parallel dynamics of electrons must be included. The time step limitations associated with the high electron thermal velocity, combined with the generally slow time evolution of wave spectra, conspire to make such calculations prohibitive in global simulation models. A flux tube simulation code based on a gyrokinetic electromagnetic model has recently been developed. Electron scale instabilities driven by the electron temperature gradient produce surprisingly high transport rates—the zonal flow dynamics, so important for ion-temperature-gradient turbulence, is absent and allows the fluctuations to grow to very large amplitude. Simulations that span the electron and ion scales but that do not address the global transport issues are now becoming feasible. Such models will, for the first time, facilitate the exploration of cross coupling between electron- and ion-scale turbulence in the plasma core.

Turbulence simulations of the colder plasma edge have been proceeding in parallel with simulations of the core. The increased collisionality associated with lower edge temperatures justifies a model based on the collisional two-fluid equations. Several codes have been under development for nearly a decade and now include a broad spectrum of physical effects: equations for the electron and ion temperatures, the density, and the electromagnetic fields, including the magnetic perturbations and associated transport.

A number of linear instabilities are described by these equations for the plasma edge, including ideal and dissipative ballooning modes and the ion-temperature-gradient modes that dominate the core. The ballooning modes are localized in the region of "bad" magnetic field line curvature. At the edge, the short scale-lengths of the ambient temperature and density profiles conspire to force the ion-temperature-gradient modes to very short wavelengths, and the resulting transport is typically smaller than from dissipative ballooning modes and drift waves. A surprise from the study of edge turbulence was the emergence of a nonlinear drift-wave instability, which dominates linear instabilities over a range of parameters at the plasma edge. For initial perturbations of very small amplitude, the turbulence simply decays away; however, above a critical amplitude the turbulence is self-sustaining. The parameters for which dissipative ballooning modes dominate drift waves and vice versa have been delineated. A second surprise was the strong impact of magnetic perturbations on edge turbulence. At high density, the magnetic perturbations dramatically increase transport, and there is some indirect evidence that this strong transport enhancement may be linked to the density limit that was discussed earlier. At low density and high temperature, the magnetic perturbations have the opposite effect. They weaken the turbulence to the extent that transport barriers spontaneously form above a threshold in β.

A complete turbulence-based simulation of the transition to the H-mode has been completed. There is good correlation between the theoretical predictions of the onset of the H-mode and observations in a number of tokamak experiments. Nevertheless, as in the case of core turbulence, a solid scientific case has not yet been made that the existing theories are a valid description of edge turbulence.

There has also recently been a substantial effort to experimentally understand and control transport in the reversed-field pinch configuration. It has been demonstrated that in the core of a reversed-field pinch, the magnetic fluctuations associated with MHD dynamo action are the primary driver of transport, and that in the edge, shorter-wavelength "electrostatic" fluctuations are the primary driver of transport. There has as yet been no substantial effort to model thermal and particle transport in these systems.

FINDINGS AND RECOMMENDATIONS

Findings

1. *The quality of the science that has been deployed in pursuit of a practical fusion power source (the fusion energy goal) is easily on a par with that displayed in other leading areas of contemporary basic and applied physical science.*
   - Learning to produce, confine, heat, and manipulate high-temperature plasma in the fusion regime is one of the most scientifically challenging endeavors in the history of mankind. The challenge is a consequence of the strong nonlinearities intrinsic to the plasma medium and the difficulty of diagnosing a medium whose temperatures greatly exceed those at the surface of the Sun—physical probes inside the plasma cannot survive the intense thermal energy fluxes. The complexity of the plasma medium is reflected in the scientific richness of the field.
   - The fusion program has been the fundamental driver for the development of the field of plasma science, including energy principles for exploring plasma stability, wave dynamics, resonant wave-particle interactions, chaos, magnetic reconnection, plasma turbulence, and transport. The techniques it uses have broadly spun off to allied areas of science (see Chapter 4).
   - While the significant scientific achievements of the fusion program are many, particularly impressive has been the development of techniques for suppressing and manipulating the small-scale turbulence that governs the leakage of stored energy from the magnetic container. Remotely controlling turbulence in a 300-million-degree medium is a premier scientific achievement by any measure and reflects the generally high quality of the science being carried out within the program.

2. *The study of high-temperature plasma has historically had a strong empirical emphasis. With the development of new theoretical, computational, and experimental capabilities, a fundamental transition away from the empirically dominated approach is now taking place.* Scientific computation, allied with theoretical insight, is becoming an integral and necessary component in the formulation and interpretation of experimental results. The process of confronting computational and theoretical results with experiment is creating new opportunities for achieving fundamental insight into the dynamics of plasma from the macroscale to the microscale.

3. *Scientific discovery has played and continues to play a critical role in shaping the direction of research and facilitating the significant enhancements in the energy containment properties of magnetic bottles that have been achieved over the history of the fusion program.* Notable examples include the following:
   - *The concept of "second stability," in which higher-pressure plasma actually becomes more stable than low-pressure plasma.* This, for example, led to the development of the spherical torus confinement configuration, which now holds the record for the highest value of β (the ratio of plasma to magnetic pressure), which is a measure of the efficiency with which the externally applied magnetic field contains the high-temperature plasma.

- *The idea that high-pressure plasma in a torus can drive its own current.* Such current, known as bootstrap current, is critical to the development of steady-state (as opposed to pulsed) confinement in tokamaks and other configurations.
- *The formation of transport barriers (which enhance containment) first in the edge of tokamaks (H-mode) and later in the plasma core and in other devices.* This discovery produced a paradigm shift: The widespread belief that energy leakage from magnetic bottles was intrinsic crumbled as increasingly sophisticated techniques for controlling small-scale turbulence were developed.

4. *Since the redirection of the fusion program in 1996 toward establishing the scientific knowledge base for fusion, there has been more emphasis on understanding the plasma dynamics underlying the operation of the various confinement configurations. However, performance goals rather than scientific overarching goals continue to be the primary driver for the allocation of resources in the program.* (See Chapter 3 for further discussion.) This emphasis is reflected, for example, in the categorization of experiments as concept exploration, proof of principle, and performance extension. All of these categories pertain strictly to the reactor potential of an experiment rather than its scientific merit—there is no parallel measure of scientific worth. The result is that experiments designed to explore an important scientific topic, even though it may be critical to the program, are at a competitive disadvantage, so that some scientific topics remain inadequately explored. Given the historical impact of scientific discovery on the program, the absence of a science-based strategic planning process is inhibiting progress. Recent efforts to rectify this deficiency are to be applauded. One example is the experimental program for DIII-D, which awards device time to thrust areas that are, for the most part, performance oriented. A parallel set of scientific topics now introduces scientific goals that cut across the various performance thrusts, raising the priority of specific scientific topics in the program. The committee is encouraged by the use of parallel scientific goals and the inclusion of outside scientists, although for the most part, time is still allocated based on performance thrusts.

5. *The U.S. program traditionally played a central role as a source of innovation and discovery for the international fusion energy effort. The program has been distinguished by its goal of understanding at a fundamental level the physical processes governing observed plasma behavior.* This feature, a strength of the program, was formalized in the 1996 restructuring, which placed a new emphasis on establishing the knowledge base for fusion energy. The quantitative detail in which experiments are designed and executed in the United States has become a benchmark for the rest of the world. The forte of the U.S. program is the confrontation of theoretical results with experimental data, along with the development of advanced computational physics codes for the quantitative exploration of novel physical concepts.

Recommendations

Increasing our scientific understanding of fusion-relevant plasmas should become a central goal of the U.S. fusion energy program on a par with the goal of developing fusion energy technology, and decision-making should reflect these dual and related goals.

Since the redirection of the fusion program in 1996, a greater emphasis has been placed on understanding the basic plasma dynamics underlying the operation of the various confinement configurations. The new emphasis on exploring scientific issues has been effectively implemented at the level of individual experiments. However, at the programmatic level, performance goals rather than overarching scientific goals continue to act as the primary driver for the allocation of resources. (See Chapter 3 for further discussion.) This emphasis is reflected, for example, in the categorization of experiments as concept exploration, proof of principle, and performance extension, which appear to measure the reactor potential of an experiment rather than its scientific merit—there is no parallel measure of scientific worth. Given the significant historical impact of scientific discovery on the program, the absence of a science-based strategic planning process is inhibiting progress.

The program direction should be determined by a focused set of scientific goals. DOE, in full consultation with the scientific community, needs to define this limited set of important scientific goals for fusion energy science. The goals should be realistic and specific so that a concrete strategy can be formulated to attack an issue and the theoretical/experimental/diagnostic tools can be marshaled. It is expected that many of the goals will transcend specific devices and therefore will serve to strengthen the linkages between different elements of the program. The committee understands that such a scientific planning process is under discussion but declines to make a judgement about the new process since it has not yet been implemented.

The achievement of scientific goals should serve as a metric for defining success within the program and should replace the previous emphasis on performance as the primary measure of progress. Improvements in scientific understanding and progress towards fusion energy are coupled, and both should serve as measures of program success and be given equal weight.

The program planning and budgetary justification carried out by DOE must be organized around answering key scientific questions as well as around progress toward the eventual energy goal. This applies to the confinement configuration program, as well as to programs of a more general nature (see Chapter 3).

Public and congressional advocacy should emphasize progress in science as well as progress toward a practical fusion power source.

Success in increasing the extent to which theory, computation, and experiment can be compared and used to validate scientific ideas will require a concerted effort. The key elements of theoretical models must be confronted with experimental observations at a level that uncovers the essential dynamics.

Essential to this effort are the following:

• An expanded effort to identify and implement the diagnostic tools required to compare experiments with theoretical models and
• The allocation of sufficient time on existing experiments to address key scientific issues or the construction of dedicated experiments to explore a specific scientific issue.

3

Plasma Confinement Configurations

INTRODUCTION

A key goal of the U.S. fusion program is to answer important physics questions (as described in Chapter 4), with the aim of acquiring predictive capability (as described in Chapter 2). A crucial ingredient in this effort is the study of a range of confinement configurations, as described in this chapter. The hallmark of laboratory physics is its ability to provide an experimental configuration optimized to reveal the physics in question. In plasma physics, many basic questions are best illuminated by studying a family of configurations, each of which is capable of stable plasma confinement. Plasma behavior is determined in large part by the spatial structure of the confining magnetic field. The experimenter can manipulate the magnetic field to vary crucial properties such as curvature, field line pitch, field strength (relative to plasma pressure), and spatial symmetry. The result is a family of related confinement configurations that can be used for the controlled study of scientific issues.

Throughout most of the history of the fusion program, a variety of confinement configurations were investigated. However, in the past each configuration was largely viewed in terms of its suitability as a fusion reactor. Progress in a configuration was judged mostly in terms of progress toward reactor-relevant plasma parameters. Now, however, the restructured program has a stronger scientific focus and the variety of configurations under study is being broadened, motivated by the following rationale. First, a range of configurations is needed to examine the array of crucial plasma science issues. Indeed, even a configuration that does not prove to be a suitable source of fusion power may be of unique value for the study of a specific scientific issue. The choices of configurations for study are being driven more than before with science as a criterion, and research is being conducted with emphasis on the scientific coupling between configurations. Second, a particular configuration can be investigated for its potential as the core of a fusion power source; that is, the set of physical attributes possessed by a particular configuration may ultimately prove advantageous for a reactor. The optimal fusion configuration is not yet known—it may be a direct extrapolation from a present concept, an evolved hybrid of concepts presently under study, or even a concept not yet articulated.

In short, a broad family of concepts bound by common physics principles may be studied both to elucidate the fundamental physics and to stimulate the scientific innovation needed for fusion energy development. The U.S. fusion program is fruitfully becoming a program that recognizes the complementary, coupled contributions of different configurations. In addition, an optimal scientific program would contain experiments spanning a diversity of scales, with small experiments being optimal for some studies and large experiments necessary for others.

In this chapter the motivation for the fusion concept program and the program's status are summarized. First, a small set of illustrative fundamental science issues is discussed, together with how they are best investigated in a variety of plasma configurations. Next, the impact of advances in these basic physics questions on the fusion energy goal is presented. The inertial confinement approach to fusion energy is also a large, active, and challenging endeavor, with important scientific underpinnings and opportunities. Although beyond the scope of this report, it, too, is discussed briefly. Another section touches on representative connections between the U.S. fusion program and the broader international program. The engineering challenges, which are discussed briefly, are also beyond the committee's charge. In the penultimate section, the metrics in place in the OFES program are discussed. Conclusions and recommendations regarding the fusion concept program are found in the last section. Appendix C contains brief descriptions of the various configurations being explored.

IMPORTANT PHYSICS QUESTIONS MOTIVATING RESEARCH WITH VARIOUS CONFIGURATIONS

The fusion research program is in part driven by a set of important physics issues that are being addressed through research using a diversity of plasma configurations. An example of one of these configurations, a tokamak, is shown in Figure 3.1. The major magnetic configurations now under study are described in detail in Appendix C. Many of these configurations are potential fusion concepts. Several of the important physics issues are briefly discussed, and for each the plasma configurations that are being employed to research the issues are identified. The issues can serve as organizing elements for the portfolio of configurations, as the committee recommends in the last section of this chapter, because the various plasma configurations under study follow naturally from them. The four physics challenges that the committee has selected are neither exhaustive nor necessarily optimal for planning purposes; rather, they are illustrative, and the actual development of a complete set of physics issues is left to the research community.

Understand the Stability Limits to Plasma Pressure

All plasma configurations possess an upper limit on the pressure (product of plasma density and temperature) beyond which the plasma becomes unstable and disassembles. In the parameter regime in which the plasma is well-described as a magnetized fluid, the theory of magnetohydrodynamics (MHD) can be used to evaluate the pressure limit for different magnetic structures and can also be used to model the detailed evolution of a plasma instability. Techniques to solve the MHD equations are well developed and can be used to accurately evaluate whether a plasma is stable to small perturbations (linear stability).

Linear instabilities can, if sufficiently global in spatial extent, cause a rapid degradation of plasma confinement. Predictions of linear instability thresholds have been tested in small-scale plasma configurations. For example, the dependence of stability on magnetic curvature has been investigated through a range of configurations with varying curvature (beginning with magnetic mirror configurations). The

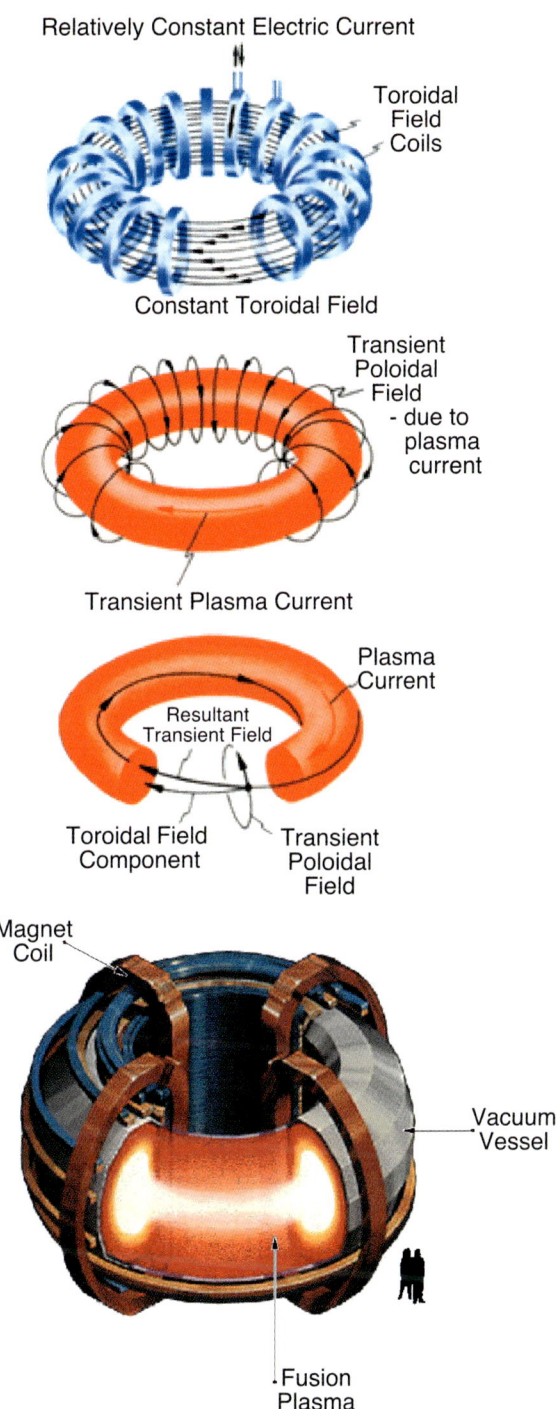

FIGURE 3.1 Components of the tokamak confinement configuration, one of the more advanced plasma confinement concepts. It uses a strong toroidal field created by external field coils (top) to stabilize the plasma while using a poloidal field created by a toroidal plasma current to confine the particles. The final configuration includes a large vacuum vessel to isolate the hot plasma from the surrounding environment (bottom). Courtesy of General Atomics and PPPL.

connection between magnetic curvature and macroscopic stability has been firmly established. More recently, the pressure limit to global instability observed in a range of tokamak experiments has been shown to agree with MHD theory.

The continuing challenge is to develop a quantitative understanding of the nonlinear evolution of instabilities, of the relationship of stability to properties of the magnetic field (such as magnetic curvature and shear), and of additional effects on stability that cannot be treated within the MHD fluid model (such as the effects of large particle orbits, separate electron and ion dynamics, and energetic particles). Moreover, there are instabilities that arise only in the presence of a finite amplitude perturbation and that would not appear within a small amplitude linear theory. At the forefront of this endeavor is the study of plasma with pressure comparable to the magnetic pressure that confines the plasma. Plasma pressure is gauged in comparison to the magnetic pressure and is characterized by the dimensionless parameter β, the ratio of the plasma pressure to magnetic pressure. Plasma pressure in the tokamak configuration has been increased over the years, so that β values of about 10 percent are now readily obtained. Extreme pressures, with β close to unity, may be obtained by pushing various magnetic field properties—for example, toroidal curvature—to an extreme. The experimental and theoretical treatment of such configurations provides a testbed for MHD stability; conversely, a challenge to our understanding of plasma stability is to develop configurations with β close to unity.

A large number of plasma configurations contribute to this goal. As one accentuates the curvature of a plasma torus (for example, by reducing the hole in the center of the torus), the curvature of the field becomes highly favorable for stability and the β limit is predicted to approach unity (Figure 3.2). This configuration, the spherical torus, provides a test of stability at extreme toroidal curvature. In the stellarator family of configurations, the magnetic field is produced largely by external coils. The magnetic field properties can be varied controllably to isolate the impact of geometry on stability. Thus, although the β value of the stellarator is relatively modest, it can provide key input to the understanding of MHD stability. In addition, stellarators can operate with minimal plasma current, allowing us to distinguish the pressure and current contributions to the free energy source that drives instabilities. The reversed-field pinch (first explored in the United Kingdom) and spheromak configurations permit study of plasmas with unfavorable curvature but strong stabilizing shear (rate of twist) in the magnetic field lines. Finally, there are two toroidal configurations, the field-reversed configuration (FRC) and the dipole, in which the field points entirely in the poloidal direction—that is, the field that is directed the long way around the torus vanishes. The FRC is predicted by MHD theory to be unstable, yet experimental FRC plasmas are stable, persisting for times much longer than the growth times of predicted instabilities. This is an example of a plasma in which effects beyond the MHD model are needed to explain experimental results.

Understand and Control Magnetic Chaos in Self-Organized Systems

In configurations in which the confining magnetic field is weak, the plasma is less stable and the magnetic field can become turbulent. The result is that the magnetic field lines can fluctuate and wander chaotically in space, leading to the random transport of particles and loss of energy. On the other hand, magnetic chaos also causes the magnetic field to rearrange spontaneously (self-organization). The physics of this laboratory process is similar to that in the spontaneous generation of magnetic field in the Earth, stars, and galaxies, a process known as the dynamo. Self-organization also occurs through the process of magnetic reconnection, which is also prevalent in solar and astrophysical plasmas.

Experimental and theoretical investigations are aimed at understanding the origin of magnetic self-organization, dynamo activity, and reconnection and the mechanisms by which the magnetic fluctuations

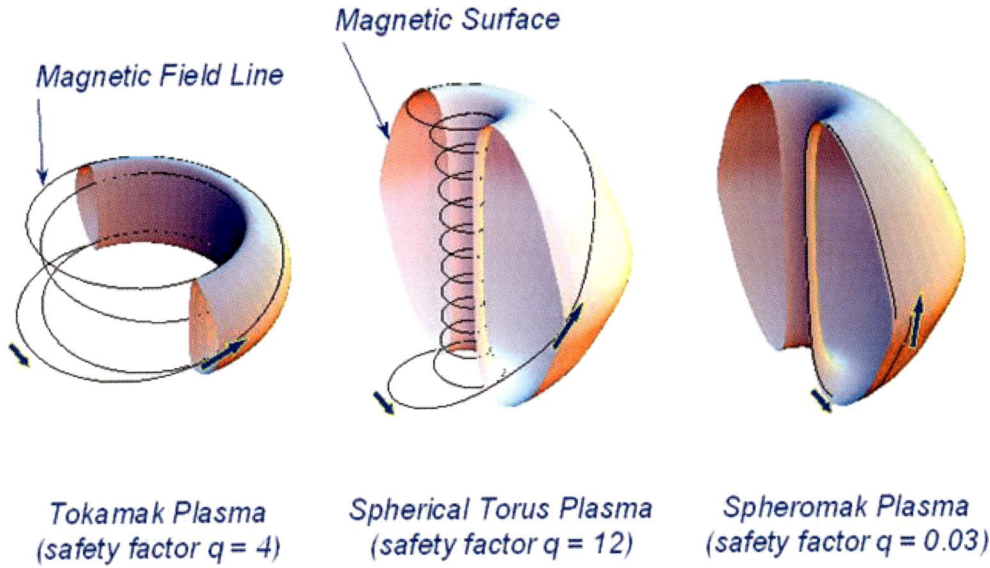

FIGURE 3.2 Examples of the magnetic topologies of several related toroidal configurations with increasing curvature and varying stability characteristics. The tokamak (left) uses a strong external toroidal field to provide robust stability against pressure- and current-driven instabilities. The spherical torus uses a weak toroidal field in a compact configuration to allow access to higher β values than obtained in the tokamak. The spheromak (right) uses internal plasma currents only to provide the confining poloidal field plus a weak toroidal field. A larger safety factor indicates a higher level of protection from current-driven instabilities. Courtesy of M. Peng (PPPL). Reprinted by permission from *Physics of Plasmas*, 2000, vol. 7, pp. 1681-1692.

drive transport. This understanding can be used to develop techniques to control the chaos and transport. Control techniques can be used to test our understanding of magnetic fluctuations and transport, as well as to enhance confinement for the fusion application. These issues are critical to, and best studied in, the class of configurations with low magnetic field, such as the reversed-field pinch (Figure 3.3) and the spheromak. Magnetic self-organization also can occur in high field configurations under special situations, such as in a tokamak during the reconnection process that underlies a sawtooth (relaxation) oscillation, as discussed in Chapter 2.

Understand Classical Plasma Behavior and Magnetic Field Symmetry

The transport of particles through the plasma arising only from collisional interactions between particles (two-particle correlations) is referred to as classical transport. The effect of instabilities and turbulence is neglected. Some aspects of plasma behavior can be dominated by classical processes, even in the presence of turbulence. The classical processes are determined, in part, by underlying spatial symmetries in the magnetic field.

To confine particle orbits, it is advantageous for the magnetic field to have a direction of symmetry. The presence of a symmetry coordinate favorably constrains the motion of particles in such "two-dimensional" systems so they do not drift out of the magnetic container. A class of confinement concepts

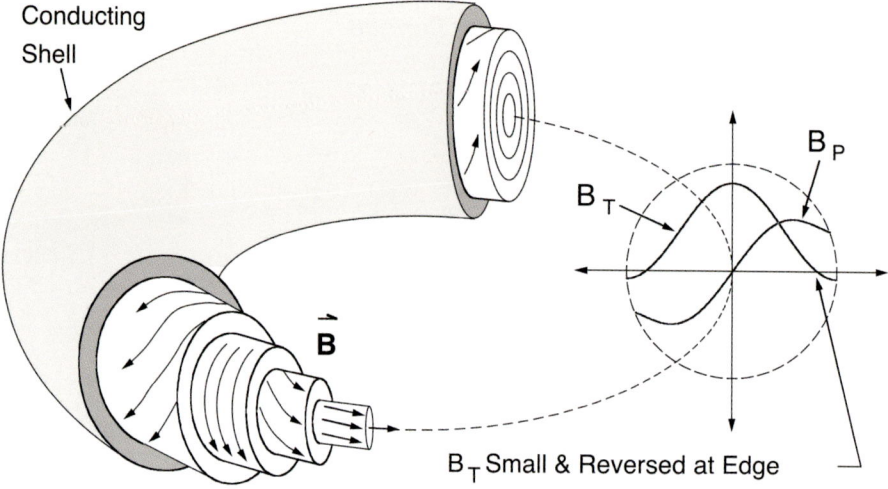

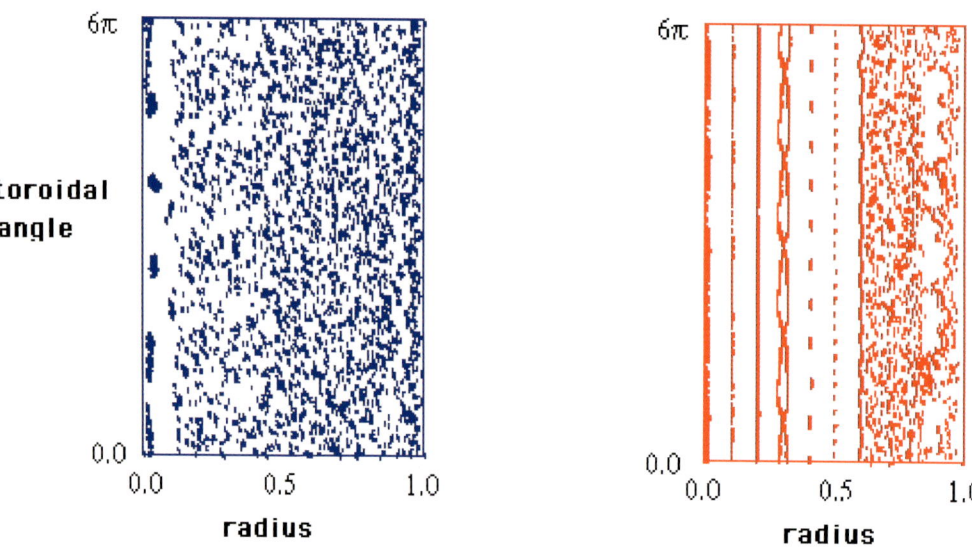

FIGURE 3.3 A magnetic confinement concept such as the reversed-field pinch (top) is a relatively self-organizing configuration that is subject to turbulent magnetic field structures. The magnetic topology includes a reversal of the toroidal field inside the plasma owing to plasma currents. Under normal inductive current drive, the magnetic field lines can readily become chaotic, as indicated by a puncture plot of the field lines as they traverse a poloidal plane (bottom left). With finer control of the plasma currents, well-defined flux surfaces are restored (bottom right). Courtesy of S. Prager (University of Wisconsin at Madison).

has been discovered by theoretical methods in which the magnetic field is fully three-dimensional (having no direction of symmetry) but appears to be nearly two-dimensional from the viewpoint of a moving particle in the plasma (see Figure 3.4). The design of such highly nonintuitive, "quasi-symmetric" systems has just recently become possible with the advent of new computational techniques. Thus, it is now feasible to study the new symmetry principle by investigating the relation of particle orbits (and associated diffusion) to magnetic field symmetry. Such configurations are within the stellarator class of magnetic containers.

In plasmas that are asymmetric in at least two coordinates, it has been discovered that an electric current can flow along the third coordinate direction in the absence of an electric field. This is a classical effect—a thermoelectric effect—in which the current is driven by a combination of a pressure gradient across the magnetic field and an effective viscosity arising from the nonuniformity of the magnetic field. The viscosity channels the electron or ion flows along the symmetry direction. A relatively complete equilibrium plasma kinetic theory exists to describe this self-generated "bootstrap current." A magnetic configuration that exhibits a very large bootstrap current, such as a toroidal plasma with a very strong poloidal asymmetry but with toroidal symmetry, is needed to test the theory and explore the robustness of the self-driven current. The spherical torus possesses these characteristics. Theory predicts that the spherical torus, when heated to high pressure, can approach the limit in which nearly all the plasma current is bootstrap-driven. The magnetic symmetry properties of a stellarator can be adjusted continuously so that experiments in this configuration offer opportunities for controlled tests of the theory.

Understand Plasmas Self-Sustained by Fusion ("Burning" Plasmas)

In all existing magnetic fusion experiments the plasma is heated by external sources. The energy input is required to overcome the inevitable energy loss from the plasma. In a plasma containing fusion reactions, a state can be reached in which the plasma is self-sustaining: the alpha particles produced in the fusion reaction deposit their energy back into the plasma at a rate sufficient to keep the plasma at a fixed temperature. In such a burning plasma, the external heating can be turned off and the plasma will undergo fusion burn continuously until the fuel is exhausted.

The presence of alpha-particle dynamics in the plasma introduces at least three physics issues: alpha-particle heating, alpha-particle transport, and alpha-particle-generated instabilities or turbulence. Alpha particles will transfer their energy to the background plasma, either classically through collisions or through more rapid processes involving turbulence or instabilities. The spatial profile of the heating by alpha particles can be different from that for externally heated plasmas and can alter the behavior (such as the transport) of the background plasma. Whether the alpha particles within the plasma are transported at the same rate as the background plasma, or whether the transport depends on the particle energy, is a key issue. For example, the large gyration orbits of energetic alpha particles cause them to sample the background plasma turbulence differently from the small-orbit particles that constitute the bulk plasma—transport properties may therefore differ from the bulk. Finally, the alpha particles represent an additional source of free energy, so they can introduce new plasma instabilities or turbulence, which can in turn affect confinement of either the alpha particles or the background plasma.

Since the alpha-particle dynamics depends on background plasma properties that are themselves influenced by the alpha particles, a burning plasma is a type of self-organized system, organized in part by the alpha-particle dynamics. Isolated aspects of alpha-particle effects can be treated theoretically. However, calculation of the dynamics of the plasma, including the coupled alpha-particle effects, is a complex, nonlinear problem whose complete solution remains undetermined.

(a) **Conventional Stellarator**

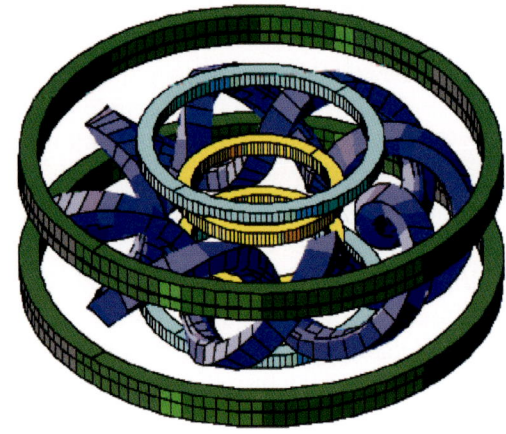

(b)

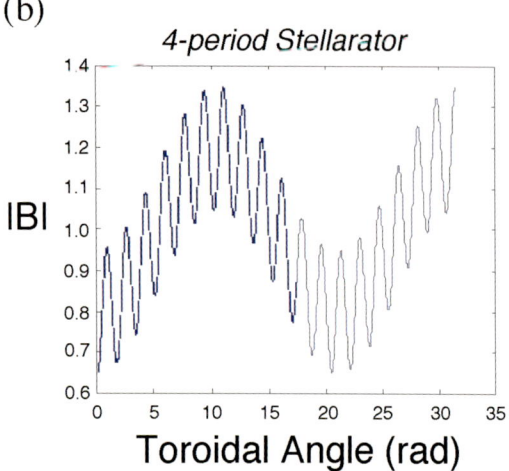

(c) **Quasi-Symmetric Stellarator**

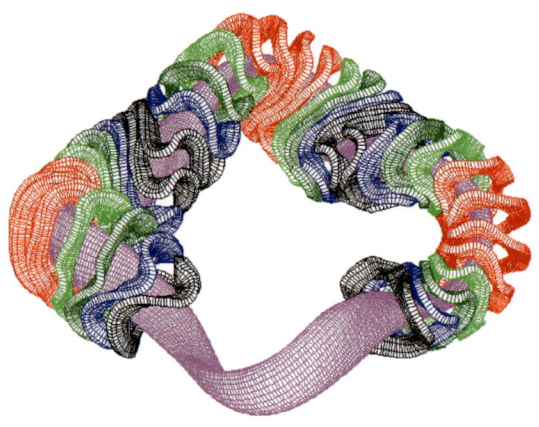

(d)

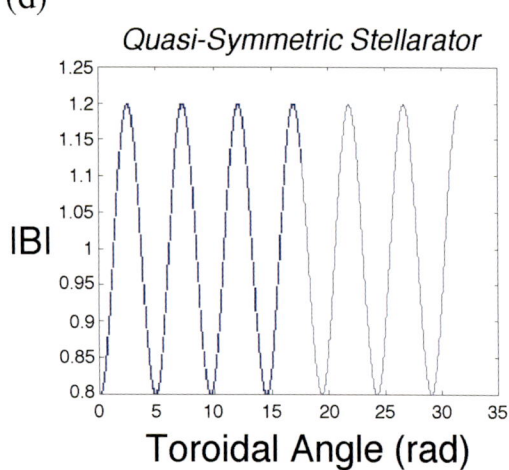

FIGURE 3.4 The stellarator concept uses complex three-dimensional coil and magnetic flux surfaces to create a quasi-symmetric configuration in which the magnetic field appears to be only two-dimensional in the frame of reference of a moving particle in the plasma. The conventional stellarator (a) has relatively simple helical symmetry and multiple harmonics in the field strength along a field line (b), which in turn gives rise to large particle losses. In contrast, the quasi-symmetric stellarator (c) eliminates the harmonics and produces a field line with single harmonic symmetry (d), effectively eliminating toroidal curvature (i.e., the long-period feature in (b)) and dramatically improving particle confinement. Courtesy of D.T. Anderson (University of Wisconsin at Madison).

Two types of experiments have been performed to investigate alpha-particle effects. First, some alpha-particle effects have been simulated in experiments through the production of energetic particles, for example by heating with radio-frequency waves. Second, experiments have been performed in weakly burning plasmas, in which the alpha-particle production rate is small but finite. Both series of experiments produced valuable information on specific alpha-particle effects but were unable to investigate the full, integrated physics of a burning plasma.

Thus, experimental investigation of a burning plasma remains a grand challenge for plasma physics and a necessary step in the development of fusion energy. The tokamak is a configuration that is sufficiently developed to provide access to a near-term burning plasma experiment, as indicated in Figure 3.5. The ITER, Fusion Ignition Research Experiment (FIRE), and Ignitor designs and other conceptual design studies of burning plasma experiments are thus necessarily based on the tokamak. The optimal route to a burning plasma experiment depends on judgements made about strategic issues such as its time urgency, the likelihood of achieving burn conditions with present designs, and the transferability of results to configurations other than that of the experiment (and, of course, the cost of the installation).

The determination of the optimal route to a burning plasma experiment is beyond the scope of the committee's charge; rather, the route should be decided in the near term by the fusion community. However, based on its interviews, as well as its observations of the Snowmass planning process, the committee believes that there is considerable evidence that the existing program is marking time on burning plasma physics issues (alpha-particle confinement and dynamics)—it is, for example, making do with paper studies until the time is ripe for the deployment of a burning experiment. The most recent planning documents available to the committee reinforce this impression. For example, the FESAC panel report[1] recommends preparation for participation in a burning plasma experiment in a 5- to 10-year time frame but gives no further specifics, and the committee found no evidence for contingency plans in case no burning experiment is undertaken by the ITER partners (Europe, Japan, and the Russian Federation). At this point, the U.S. fusion program is not demonstrating leadership in this area, apparently because Congress does not want to consider a burning plasma experiment at this time (given the history of U.S. participation in ITER, this is perhaps understandable). However, an optimal fusion science program needs two components: (1) experiments in nonburning plasmas to explore the large range of critical science issues that do not require a burning plasma and (2) experiments in burning plasmas. The first component, which would attack a broad range of issues, should not be sacrificed for the second component and can lead to scientific progress in the absence of the second component. Nevertheless, it is clear from the discussion of physics issues in Chapter 2 and in this chapter that a burning plasma experiment is required to address key issues that cannot be fully explored by the present portfolio of experiments. Thus, the United States must soon explore options for pursuing alpha-particle physics issues, possibly as a part of an international team.

REACTOR DESIGN FEATURES MOTIVATING FUSION CONCEPT DEVELOPMENT

In the last section, physics questions were discussed that motivate research employing a variety of plasma configurations and that can serve as organizing elements for a program of multiple confinement configurations. Although it may be obvious that a fundamental understanding of plasma behavior in

---

[1]Department of Energy (DOE), Fusion Energy Sciences Advisory Committee, Panel on Priorities and Balance. 1999. *Report of the FESAC Panel on Priorities and Balance*. Washington, D.C.: DOE.

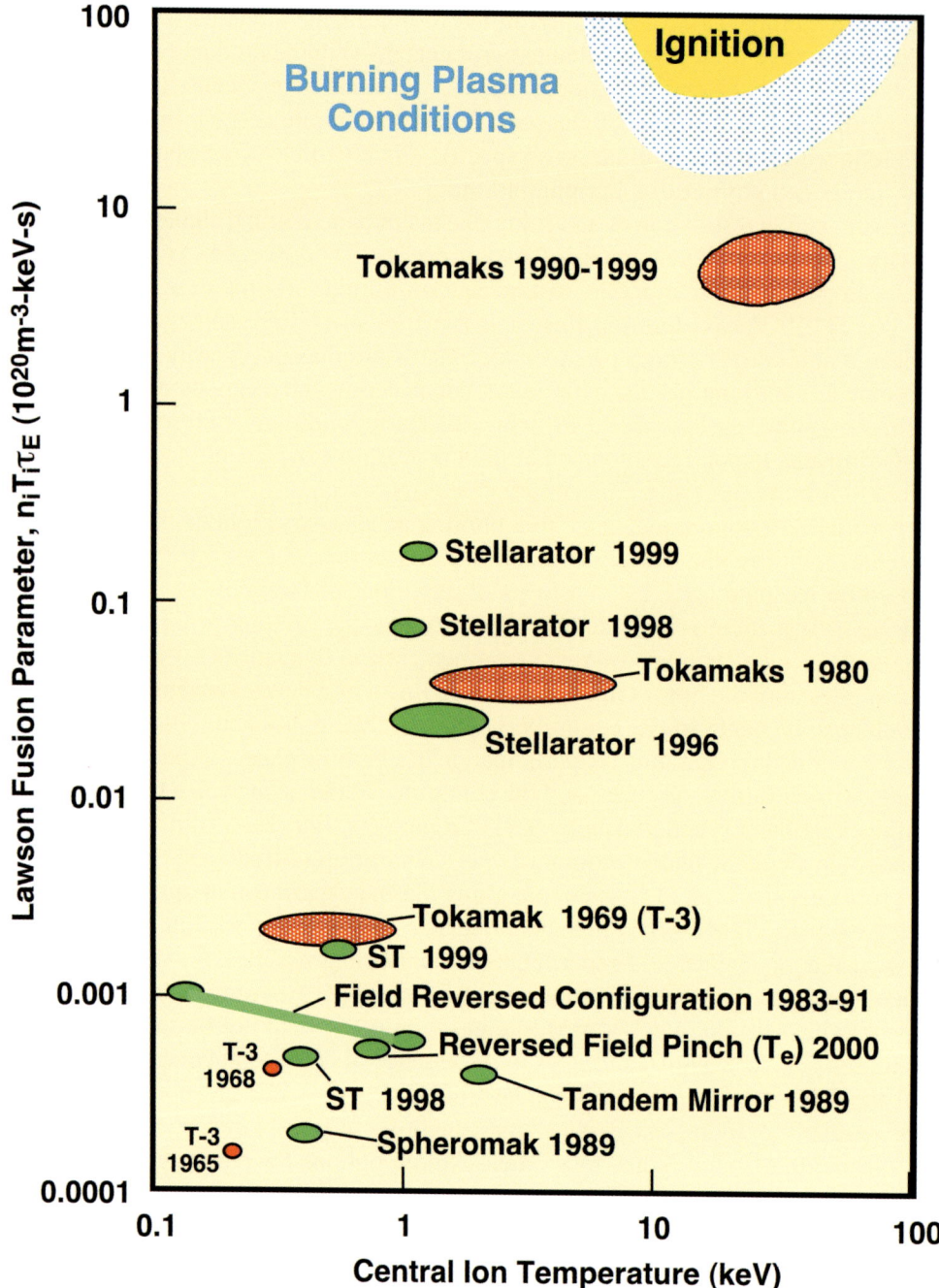

FIGURE 3.5 Improvements in plasma stability and confinement obtained in magnetic confinement configurations should allow the study of burning fusion plasmas in the near future. The Lawson fusion parameter is the product of the plasma density, ion temperature, and average confinement time and represents a simple figure of merit for proximity to conditions for fusion ignition. Courtesy of PPPL.

magnetic fields will lay a foundation for progress in fusion energy, this section briefly describes the connection between the basic physics goals of the preceding section and the development of a fusion system with attractive design features (such as small size, simple magnets, continuous operation, absence of sudden terminations, and low recirculating power):

- *Understand the stability limits to plasma pressure.* The fusion reaction rate of a plasma is proportional to p^2, where p is the plasma pressure. Hence, the higher the plasma pressure (or β), the more attractive, generally speaking, is a fusion energy system. For example, a plasma configuration with a high β limit can operate with either a weaker magnetic field (simplifying the magnet requirements) or a smaller size.
- *Understand and control magnetic chaos in magnetically self-organized systems.* The class of configurations with weak toroidal magnetic field suffers from large energy transport generated by the processes associated with magnetic self-organization. If the transport can be controlled, then the reactor advantage of the weak magnetic field requirement may be realized. The development of an understanding of magnetic self-organization may also clarify the causes of the sudden termination of the plasma, which plagues some current-carrying plasma configurations. The disruption in a tokamak is the most prominent example. Since disruptions can do substantial impulsive damage to a fusion energy system, their elimination is highly desirable.
- *Understand classical plasma behavior and the role of magnetic field symmetry.* The magnetic field within plasmas that lack symmetry can be produced without externally driven current. Such plasmas can, accordingly, form the basis for steady-state operation, a desirable feature for a fusion energy system. In addition, they may be less susceptible to disruptions. Similarly, systems in which the magnetic field structure has been designed to optimize the self-driven bootstrap current can also operate in steady state with minimal need for external current drive (which then minimizes the recirculating power).
- *Understand plasmas self-sustained by fusion.* Since the presence of alpha particles can affect nearly all aspects of plasma behavior, their dynamics can have a very large impact on the realization of a fusion reactor. The influence of alpha particles on plasma transport and their effectiveness in heating the plasma directly affect the size of the plasma core of reactor. Moreover, if they were very poorly confined or if they greatly enhanced transport, they could render a confinement configuration unfeasible for fusion energy production.

INERTIAL FUSION ENERGY CONCEPT DEVELOPMENT

Inertial confinement offers an entirely distinct approach to fusion energy—one with its own set of scientific and engineering challenges as well as scientific opportunities. An inertial fusion energy (IFE) system will have three components: the target, the driver, and the fusion chamber. Just as with magnetic fusion energy (MFE) systems, there are a variety of different concepts under study to serve as implosion drivers (e.g., direct versus indirect drive) and high-repetition-rate drivers (e.g., heavy ion beams, solid state lasers, krypton fluoride lasers) and for the fusion chamber (e.g., solid walls, liquid walls). The physics issues associated with the target dynamics are immensely challenging; they include hydrodynamic stability, the equation of state of dense matter, radiation transport, and the laser-plasma interaction.

At the present time, nearly all the research associated with inertial confinement is supported by DOE Defense Programs (DP). It is particularly noteworthy that in the coming years, burn propagation physics is planned to be studied in the National Ignition Facility. In addition to developing a knowledge of the target physics, inertial fusion energy will require the development of high-repetition-rate lasers, inexpensive target fabrication techniques, and suitable confinement chambers. The Office of Fusion

Energy Sciences presently supports the development of heavy-ion-beam drivers. While the involvement of OFES in IFE development is growing, strongly leveraged by the very large DP effort, the OFES program is still heavily weighted toward MFE. Since OFES is the sole steward of MFE, this situation will probably persist for the foreseeable future. Accordingly, the committee was constituted to focus its effort on science within the MFE program, so it leaves treatment of the rich science under way within the IFE effort to another committee.

LINKAGES WITH INTERNATIONAL PROGRAMS

The dollar total of international research programs in fusion greatly exceeds the annual U.S. budget in this area. In addition, in recent years both the European Union and Japan have been more willing than the United States to commit themselves to major experimental projects that will advance the fusion effort. A major stellarator experiment (the Large Helical Device) has just come on line in Japan, and a machine of comparable scale is being constructed in Germany. The JET tokamak experiment, which is presently the only machine in the world able to operate with a mixture of deuterium and tritium and therefore to explore processes involving energetic alpha particles produced during fusion reactions, continues to operate in the United Kingdom under funding provided by the European Union. There are recent indications from Japan that the JT60-U facility may be upgraded. No facilities of a comparable scale exist or are close to approval in the United States. As discussed in the introduction to Chapter 2, however, the United States, because of its investment in diagnostics and theory and computation, continues to play a leadership role in developing the science base for fusion by linking experimental observations with a fundamental understanding of the physical processes controlling the plasma dynamics. As a consequence, the impact of the United States program exceeds what the budget figures alone would suggest.

There have long been and continue to be extensive collaborations at all levels between international participants in the quest for controlled fusion. For example, the United States provided the neutral beams used to reach the record values of the plasma β in the spherical torus experiment at the Culham Laboratory in the United Kingdom. Given the strength of the international program and the impressive facilities that are available, it is essential that the United States maintain these collaborations. The expertise in diagnostics, theory, and computation from the United States is especially suitable for maximizing the return on these foreign experiments, and our country should continue to promote these collaborations as beneficial to all parties.

To develop an understanding of the impact of alpha-particle dynamics in regimes with dominant alpha-particle heating, it will be essential to collaborate with international programs because of the high costs of constructing devices that can reach the parameters required and can handle the radiation problems safely. Should scientists in Europe decide to conduct further deuterium-tritium experiments, the United States should seek to participate as appropriate to maximize the physics return on the experiments. As mentioned earlier, the United States should begin to reestablish its international leadership role by defining an affordable burning plasma experiment that could be constructed with financial contributions from several international partners.

ENABLING TECHNOLOGIES FOR PLASMA CONFIGURATION DEVELOPMENT

This report focuses on the scientific aspects of fusion concept development. However, significant advances in engineering sciences are also critical. Two types of engineering advances are necessary—those that enable plasma experiments and those that enable a fusion power system. Often these needs overlap. Examples of the former type include the development of neutral beams to heat plasmas, radio-frequency sources for current drive and heating, pellet injectors for plasma fueling, and high field magnets for plasma confinement.

Numerous engineering advances are needed to realize fusion power. Materials research is a major challenge: a fusion reactor must be made of materials that can, among other things, withstand intense heat fluxes, intense neutron fluxes, and acceptable tritium breeding. Ideas are under investigation for new alloys and for flowing liquid walls. Other challenges include the development of fueling techniques, high field magnets, low activation materials, and remote maintenance techniques. In addition, as the physics of the concepts described in the section on reactor design features moves forward, engineering constraints become the limiting factor. For example, concepts aimed at developing compact plasmas face the physics hurdle of achieving good confinement. However, if that physics goal is achieved, the burden shifts toward the development of materials that can sustain the intense neutron bombardment, a necessary consequence of compact systems. The enormous engineering challenges of fusion power, and their contributions to engineering science, are beyond the scope of this report.

CURRENT METRICS FOR FUSION CONCEPT DEVELOPMENT

FESAC has defined three stages for the experimental development of a fusion concept, beginning with the concept exploration stage (initial experiments to investigate, at a small scale, isolated physics features of a concept), the proof-of-principle stage (with medium-sized experiments aimed at investigating the broad range of key physics issues), and the performance extension stage (where experimental parameters are brought closer to conditions in a reactor).[2] The performance extension stage will be followed by a burning plasma experiment. As this report is being written, in the United States the tokamak is being explored in facilities at the performance extension stage. At the proof-of-principle stage, only the spherical torus is under full investigation. The reversed-field pinch has been recommended by FESAC to proceed to the proof-of-principle stage; reversed-field-pinch research is now transitioning from the concept exploration stage to the proof-of-principle stage. Several configurations are under experimental investigation at the concept exploration stage, including the quasi-symmetric stellarator, spherical torus, spheromak, field-reversed configuration, magnetized target fusion, dipole configuration, electrostatic confinement, and other emerging concepts. This is an evolving set of configurations, some of which will graduate to the next level and others of which will terminate as new results unfold. These various concepts are described in Appendix C, while Figure 3.6 shows examples of experimental facilities at each stage of concept development.

In 1999, a FESAC subpanel was convened to establish criteria, goals, and metrics for the fusion program.[3] It discussed two forms of metrics: those to judge whether a fusion concept is ready to move to the next stage of development and those to judge whether the overall fusion program is properly balanced. For the former judgements, 10 criteria were described: the quality of the research, the confidence for the next step, the plasma science and technology benefit, the issue resolution capabilities, the degree to which the research is at the cutting edge in its area, the energy vision of the concept, the programmatic issues of the proposed work (cost, adequacy of resources, etc.), the influence of the

---

[2]Department of Energy (DOE), Fusion Energy Sciences Advisory Committee, Alternate Concepts Review Panel. 1996. *Alternative Concepts: A Report to the Fusion Energy Sciences Advisory Committee.* Washington, D.C.: DOE.

[3]Department of Energy (DOE), Fusion Energy Sciences Advisory Committee, Panel on Criteria, Goals, and Metrics. 1999. *Report of the Panel on Criteria, Goals, and Metrics.* Washington, D.C.: DOE.

FIGURE 3.6 Examples of the stages of experimental development of plasma configuration concepts: (a) a large advanced tokamak experiment (DIII-D) at the performance extension stage exploring plasma parameters approaching those of a reactor; (b) a mid-sized reversed-field pinch experiment exploring a range of issues at the proof-of-principle stage; and (c) a smaller spherical torus dedicated to exploring a particular type of current drive in a concept exploration experiment. Courtesy of (a) General Atomics, (b) S. Prager (University of Wisconsin at Madison), and (c) T. Jarboe (University of Washington).

proposal on the overall fusion concept research portfolio, the general science and technology benefit, and the adequacy of milestones. These criteria are meant to be applied to proposals for research at each stage of development—concept exploration, proof of principle, performance extension, and fusion energy development. However, the weighting of the criteria varies from stage to stage. For example, the energy vision gains in importance at the later stages, while contributions to science in general decrease in importance at and beyond the performance extension stage.

The FESAC report also discusses the need for balance in the distribution of research among the various stages of development and in the coverage of the various scientific elements. The number of research programs at the earlier stages of development is expected to be large and then to decrease steadily from stage to stage. However, since the cost of an experiment increases with its stage of development, the funding would probably be weighted toward the more expensive, advanced stages of development—even in a program with a large base of small experiments. The report also articulates the various program elements that should be active in a well-balanced scientific program. The program elements included plasma science and technology, confinement physics, confinement configurations, fusion technology, and systems analysis. General metrics for each program element are described. A continuing peer review process carried out by panels is also advocated to allow rebalancing the program as needed.

In addition to the FESAC metrics described above, ongoing systems analyses of fusion power plants serve to assess the probability that specific confinement approaches will be achieved in the plant's reactors. These system analyses incorporate a conceptual design for a fusion power plant based on assumptions about the physics and technology. Usually these assumptions are intended to be judicious extrapolations from present knowledge, so the systems studies can be best executed for the most highly developed fusion concepts. For emerging concepts that are not well understood, a complete power plant study is more speculative.

Such system studies serve to provide insight into the different fusion concepts embodied by the reactors, to reveal which physics and technology are most likely to increase the attractiveness of a fusion power plant, to provide a basis for comparing different fusion concepts, and to evaluate the role of fusion within the full portfolio of energy sources. The results of such studies are often encapsulated in a single number, the cost of electricity. However, such a simplification, which involves many uncertainties in economics and other areas, is inherently inaccurate. The more enduring value of systems studies is the guidance they provide for research.

The above categorization of the various stages of reactor development (concept exploration, proof of principle, performance extension) has been useful in establishing a program to systematically evaluate innovative concepts. It has also provided a framework for peer review. However, the categories mainly relate to the progress of individual confinement concepts toward a fusion power reactor and not to progress on understanding fundamental, cross-cutting science issues. Thus, alongside these FESAC categories of concept development, a parallel set of scientific questions should be developed, as was proposed earlier in this chapter. This would give more weight to the scientific contributions from experiments and would more openly allow for experiments in configurations not suitable for a reactor that would advance important fusion science issues.

FINDINGS AND RECOMMENDATIONS

Findings

1. *A fusion research program must investigate a range of confinement approaches.* Such a wide-ranging program would allow fusion science issues to be examined in ways not possible in a single configuration. It would also allow developing the optimal configuration for fusion energy application.

2. *The fusion program benefits from experiments covering a range of scales.* Some issues are best addressed at small scale, some at large scale. In the past, fundamental discoveries have emerged from both small and large experiments.

3. *In the past several years, the OFES program has effectively broadened the spectrum of confinement configurations under study.* At least six new experiments in nontokamak configurations have been initiated at the concept exploration level, and two have been initiated at the proof-of-principle level. In just the past year, several additional concept exploration experiments have been initiated.

4. FESAC has defined three stages for the experimental development of individual fusion concepts toward the fusion energy goal, along with metrics to assess whether a particular concept is ready to advance to the next stage of development. These categories of progress and metrics have been employed in the peer review process, which has led to the present program of innovative concepts. *While the categories are effective in assessing the progress of a given experimental concept toward the fusion energy goal, they are not effective in defining or promoting the solution of essential cross-cutting science issues.* Given that most of the concepts supported by the program will not, ultimately, manage to achieve reactor capability, this categorization underplays the broader scientific importance that such experiments could have for the program.

Recommendations

The confinement configuration program should be specified in terms of scientific questions.

The primary contribution of exploring a variety of confinement configurations is the elucidation and discovery of plasma science relevant to fusion. An individual scientific question or issue may warrant experimental investigation in a variety of plasma configurations. The scientific question should determine the needed experimental scale. Some issues are best investigated at small scale, others require a large scale.

Alongside the FESAC categories for stage of concept development, a parallel set of categories should be developed to assess how the research in the program is organized in terms of science issues. The OFES budget should also be justified according to these thematic scientific categories.

The present set of categories (concept exploration, proof of principle, performance extension) describes the progress of individual fusion concepts towards a fusion power reactor but does not reflect progress in cross-cutting scientific issues. Whereas criteria for assessing concept development and the peer review process strongly emphasize scientific contributions, the above categories do not sufficiently reflect such contributions.

A roadmap for the fusion program should be drawn up that shows the path to answering the major scientific questions, as well as the progress so far in the development of fusion concepts.

The development of a roadmap for a fusion-based energy source is essential to aid in the long-term planning of the fusion program. The roadmap should show the important scientific questions, the evolution of confinement configurations, the relation between these two features, and their relation to the fusion energy goal.

Solid support should be developed within the broad scientific community for U.S. investment in a fusion burning experiment.

Such an experiment is scientifically necessary and is also on the critical path to fusion energy. Determining the optimal route to a burning plasma experiment is beyond the scope of the committee's charge; rather, the route should be decided in the near future by the fusion community. Resources above and beyond those for the present program will be required. The U.S. scientific community needs to take the lead in articulating the goals of an achievable, cost-effective scientific burning experiment and to develop flexible strategies to achieve it, including international collaboration.

The committee agrees with the SEAB report that "... development both of understanding of a significant new project and of solid support for it throughout the political system is essential."[4] However, since the U.S. fusion energy effort is now positioned strategically as a science program, advocacy by the larger scientific community for the next U.S. investments in a fusion burning experiment now becomes even more critical to developing that support. For this reason alone, the scientific isolation of the fusion science community needs to be lessened.

There should be continuing broad assessments of the outlook for fusion energy and periodic external reviews of fusion energy science.

A planned sequence of independent external reviews should replace the current pattern of multiple program reviews of different provenance (e.g., this review and recent SEAB and FESAC reviews). These reviews should be open, independent, and independently managed. They should involve a cross section of scientists from inside and outside the fusion energy program. The manifest independence of the review process will help ensure the credibility of the reviews in the eyes of Congress, OMB, and the broader scientific community.

The scientific, engineering, economic, and environmental outlook for fusion energy should be assessed every 10 years or so in a process that draws on the expertise of fusion scientists, other scientists, engineers, policy planners, environmental experts, and economists, from the United States and elsewhere. The assessment should examine from multiple perspectives the progress in the critical interplay between fusion science and engineering in light of the evolving technological, economic, and social contexts for fusion energy.

Consonant with its charge, the committee has not taken up the many critical-path issues associated with basic technology development for fusion or the engineering of fusion energy devices and power plants, yet it is the combined progress made in science and engineering that will determine the pace of advancement toward the energy goal. Moreover, much of fusion science research is undertaken in the expectation that it will contribute to the energy goal. Regular, formal assessment of the progress towards fusion energy is one important way in which a fusion science program can be made accountable to the long-range energy goal.

---

[4]Department of Energy (DOE), Secretary of Energy Advisory Board, Task Force on Fusion Energy. 1999. *Realizing the Promise of Fusion Energy: Final Report of the Task Force on Fusion Energy.* Washington, D.C.: DOE, p. 2.

4

Interactions of the Fusion Program with Allied Areas of Science and Technology

INTRODUCTION

Historically, the development of the fusion program involved both basic physics and the applied and engineering sciences. However, because the energy goal of the fusion program is an application, the contributions of this program to our understanding of fundamental physics are sometimes obscured. In reality, high-temperature plasmas are not only of great intrinsic scientific interest but also of great general interest in fields from astrophysics to material science, with the goal of basic plasma physics being to elucidate the linear and nonlinear properties of the plasma medium under the wide variety of conditions in which it is encountered in nature and in the laboratory. The fusion program should be credited with having played a major role in advancing our experimental and theoretical understanding of this "fourth state of matter."

Many of the fundamental concepts of fusion science, ranging from linear stability theory to kinetic theory and nonlinear dynamics, as well as its experimental techniques, have close connections to other areas of physics. Plasma physics has, in fact, been the incubator for a number of key research areas in modern physics, and in some cases, plasma scientists became leaders of the emerging fields—solitons, chaos, and stochasticity are noteworthy examples. In addition, basic tools developed in the fusion program, from computer-based algebra to particle simulation techniques, have found widespread application in allied fields.

Historically the United States has been a leader in advancing all aspects of magnetic fusion and plasma science, particularly with respect to the ability of its fusion program to pose and answer deep scientific questions on issues relevant to both fusion and the broader scientific world. At a time when U.S. funding for the fusion effort has been cut back, it is important to ask what the forefront topics of interdisciplinary research are, whether the program is continuing to engage these critical topics, and whether it is advancing the computer and technological tools required to solve the critical problems facing plasma and fusion science. In the following sections, the committee discusses a number of these issues.

WHAT ARE THE DEEP SCIENTIFIC CONTRIBUTIONS THAT HAVE IMPACTED OTHER PHYSICS FIELDS?

The core objective of the fusion energy science program is to reach a fundamental physical understanding of the behavior of high-temperature plasmas in the context of plasma configurations capable of plasma confinement sufficient for economic energy extraction. Fusion science aims to study the stability properties and transport behavior of such systems, and in order to conduct these studies, it has made progress in a number of topical areas that have had a broad impact on the larger scientific and industrial community. Examples of some cross-cutting research topics are stability theory; stochasticity, chaos, and nonlinear dynamics; dissipation of magnetic fields; origins of magnetic fields; wave dynamics; and turbulent transport.

Stability Theory

The complex plasma dynamics due to macroinstabilities observed in early plasma experiments motivated the development of powerful energy principles and eigenmode techniques for exploring the linear stability of plasma equilibria. The wide variety of instabilities in plasmas, which span an enormous range of spatial and temporal scales, defines the richness of the plasma medium and challenges us to understand its dynamics. Past research supported by the fusion program greatly improved our ability to predict the thermal pressure beyond which a plasma will disassemble. These predictions were confirmed in experiments in which the plasma temperatures exceeded those found in the core of the Sun. Experimental explorations led to methods that significantly increase the plasma pressure limits set by stability. It speaks to the quality of those studies that the stability analysis techniques developed by the fusion program—such as the energy principle, the notion of convective instability, and weakly nonlinear stability theory—are now essential tools not only in the field of plasma science but also in allied fields such as fluid dynamics, astrophysics, and solar, space, ionospheric, and magnetospheric physics.

Stochasticity, Chaos, and Nonlinear Dynamics

Understanding how magnetic field topology—the existence of bounding magnetic flux surfaces that lead to hot plasma confinement—is controlled by both local and global physical processes and how such bounding flux surfaces break up is a critical research topic for fusion. Interestingly, the same kind of issues also emerged in the description of how an essentially collisionless, unmagnetized plasma is heated; there, the onset of stochasticity needed to be understood in velocity space. This research followed on the heels of the pioneering Kolmogorov-Arnold-Moser (KAM) description of the onset of chaos and therefore contributed to the rapid progress in this field. A number of fundamental tools, including the standard map—now in common use in studies of the onset of stochasticity and chaos in far more general physical settings—came from plasma scientists. Studies of nonlinear wave-wave and wave-particle interactions relevant to both plasma confinement and transport played an important role in the development of tools for dynamical systems theory and nonlinear dynamics, and it was senior scientists trained in the physics of plasmas who developed the first published method for controlling chaos.

Dissipation of Magnetic Fields

A fundamental physics challenge has been to explain the observed very short timescales that characterize the release of magnetic energy in the solar corona, in planetary magnetospheres (including that of Earth), and in fusion experiments. Classical collisional dissipative processes are orders of magnitude too weak to explain the timescales observed. The difficulty lies in the extreme range of spatial scales, from the macroscopic to the microscales associated with kinetic boundary layers, and in the necessity to include kinetic processes to provide collisionless dissipation. An emerging understanding based on theory, computation, and basic experiments is linked to the mediating role of dispersive waves, which act at the small scales where the frozen-in condition is broken. For the first time, the predictions of energy release rates in fusion experiments are consistent with observations. One consequence of the fast release of magnetic energy associated with magnetic reconnection in some fusion experiments is the evolution to a minimum energy state (the Taylor state, named after the U.K. physicist who made the theoretical prediction), where the magnetic field is partially self-generated by the plasma. The resulting "dynamo" action is related to magnetic dynamo processes in astrophysical systems such as the Sun, the planets, and the galaxy, discussed next.

Origins of Magnetic Fields

The origin of cosmic magnetic fields is another of the fundamental puzzles of physics. Magnetic fields are ubiquitous in nature. Objects as diverse as planets, stars, galaxies, and galaxy clusters show evidence of magnetic fields and show phenomena that are intimately tied to the presence of these fields, but—with the possible exception of Earth's planetary dynamo—their origin is not understood. Although theory and simulations have made great strides in elucidating the basic physical principles of magnetic field generation, the key missing piece has been controlled laboratory experimentation. With a few exceptions, astrophysicists, geophysicists, and applied mathematicians led the most important past developments in dynamo theory and simulations; now, with the recognition that fast dynamo processes also play an important role in fusion devices, fusion researchers have begun to play an important role as well. On the experimental front, the dynamo in the reversed-field pinch device remains among the few laboratory demonstrations of a turbulent dynamo. These fusion experiments are pushing the subject into previously unexplored areas, including the development of magnetic dynamo theory beyond the single-fluid MHD regime. These studies will be of direct relevance to the origins of magnetic fields on the largest scales in the universe, seen in galaxies and clusters of galaxies.

Wave Dynamics

The plasma state is unique in the rich variety of waves that are supported by it. Waves in plasmas not only appear spontaneously as a consequence of instabilities but also can be generated to control plasma temperature and currents. Understanding how waves propagate and are absorbed in nearly collisionless plasma was a key scientific goal for the fusion program and has had an important impact on understanding phenomena in space plasma physics. Building on Landau's idea of the wave-particle resonance as a mechanism for collisionless dissipation, fusion scientists developed models to describe the absorption of high-power radio-frequency waves from kilohertz to multigigahertz and benchmarked the predictions in fusion experiments. Waves could then be used to engineer the phase space of particle distribution functions. Waves can now be excited in plasmas to generate intense currents or to accelerate particles to high energies, a technique that may be applied to a future generation of high-energy accelerators. The nonlinear behavior of waves has also been an intrinsic component of the science of plasma wave dynamics, and this knowledge has spread widely to many other branches of physics. Indeed, such ubiquitous concepts as absolute and convective instabilities, solitons (nonlinear waves that persist through collisions), and parametric instabilities were extensively developed within the fusion context. Important industrial applications include the use of radio-frequency technologies for plasma processing in semiconductor manufacturing. These ideas emerging from the fusion science program have also had an impact in less obviously allied science areas: for example, plasma physicists introduced the idea of using solitons for commercial high-speed communications.

Turbulent Transport

Understanding transport driven by turbulence is critical to solving such important problems as the accretion of matter into black holes, energy transport in stellar convection zones, and energy confinement in fusion experiments. Both experiment and theory have shown that gradients in pressure, angular momentum, or other sources of free energy drive small-scale turbulent flows that act to relax the gradients. This "anomalous transport" process is to be contrasted with classical transport, which arises from two-particle Coulomb interactions in magnetic fusion plasmas (as well as from photon diffusion in some astrophysical systems such as stellar interiors). The identification of anomalous transport in fusion experiments (and the corresponding theoretical work) sparked the recognition of its importance in space science and astrophysics, fields in which concepts such as anomalous transport, heat flux inhibition, and turbulent heating are now common language. This cross-fertilization continues: experimental work in fusion has shown that turbulence can be spontaneously suppressed and a transport barrier formed, and that the responsible mechanism is linked to the development of local zonal flows, which shred the vortices driving transport. These experimental developments, together with theory and simulations, have intimate connections with work on zonal flows in laboratory fluid dynamics (e.g., the Couette flow) and planetary physics, in which similar transport barriers and processes have been investigated.

WHAT HAVE BEEN THE FUSION-SPECIFIC CONTRIBUTIONS OF THE UNITED STATES TO THE WORLD PROGRAM?

In addition to the examples of fundamental science issues discussed in the previous section, a number of significant science discoveries closely linked with fusion can be identified. The United States has traditionally played a central role as a source of innovation and has to a greater extent than other countries sought to understand at a fundamental level the physical processes governing observed plasma behavior. This role is reflected in the U.S. dominance of plasma theory, which is an essential tool for unraveling the complexities of plasma dynamics.

The U.S. fusion program has advanced the world effort in a number of key scientific areas: In the late 1950s, the basic energy principle for MHD stability was largely derived in the United States, particularly using the powerful Lagrangian variational δW approach. Also in the late 1950s, U.S. scientists developed techniques for modeling collisions by Coulomb forces in a way suitable for large-scale computations and showed the existence of a large variety of kinetic waves in a magnetic field (now known as Bernstein waves). During that same period, U.S. scientists invented important magnetic confinement schemes such as the stellarator. Later, the favorable properties of the spherical torus configuration were demonstrated theoretically (and have been confirmed in recent experiments in the United Kingdom). In the 1960s, the United States also led the effort to understand resistive instabilities in magnetically confined plasmas, which led to the identification of modes, such as the tearing mode, that grow on a timescale intermediate between the MHD and resistive timescales. Resistive-MHD theory and computer simulation of tokamak instabilities in the 1970s were able to reproduce the sawtooth phenomenon and also many features of tokamak disruptions. The phenomenon of "second stability," in which increasing plasma pressure surprisingly leads to greater stability, was a discovery of the U.S. program.

Also in the 1970s, the United States led the effort to develop a variety of methods for the radio-frequency heating of plasmas to thermonuclear temperatures. In the late 1970s and early 1980s, U.S.-invented methods for sustaining plasma currents by waves stimulated noninductive current drive programs on essentially all tokamak facilities worldwide. In the 1980s, discrete Alfvén-wave instabilities in tokamaks, including those that could be driven by alpha particles, were identified, with the United States playing a leading role; these modes were first observed on U.S. tokamaks in the 1990s.

U.S. experiments first identified the ion-temperature-gradient instability as a driver for thermal transport in toroidal systems. The understanding of the role of sheared flow in reducing transport was also led by U.S. scientists. In the 1990s, the United States made the first observations of the so-called "reversed shear" mode of enhanced confinement in tokamaks.

The United States also played a leadership role in applying modern computational tools to plasma problems, including particle-in-cell methods and gyrokinetic and gyrofluid simulations. The application of these computational tools, together with theoretical advances in the understanding of turbulence and associated anomalous transport, has led to the beginnings of predictive capability for energy storage in magnetic containers, as described in Chapter 2.

WHAT ARE THE FUTURE FOREFRONT AREAS OF INTERDISCIPLINARY RESEARCH?

A wealth of exciting new areas of research can be identified for high-temperature plasma physics. Some examples primarily of interest to fusion science have already been given. Other areas include the following:

- *New methodologies for advanced diagnostics and modeling.* Fusion research is among the most advanced disciplines in terms of the integration of experiments, diagnostics, theory, and numerical simulations. Advances in this area will have significant impact on allied disciplines, including astrophysics and space physics.
- *Material science.* The development of materials that can withstand the huge energy and neutron flux of a fusion environment, as well as the large energy flux of present experiments, is an enormous challenge. The material must survive for long periods, release minimal impurities into the plasma, and be minimally radioactive. The development of new composite materials is under way, as are new techniques such as flowing liquid walls.

In many areas, the richness of new research areas relates to the deep connections between plasma physics and other physical science disciplines and complements work done purely in the fusion domain. Promising interdisciplinary research areas include the following:

- *Nonneutral plasmas.* The acceleration of energetic dense beams of ions or electrons involves the physics of nonneutral plasmas, where a number of collective effects play important roles. Techniques developed in plasma physics have played—and continue to play—a central role in the design and optimization of particle accelerators and colliders, from the heavy-ion accelerators used for nuclear studies, to the electron accelerators used as synchrotron radiation sources for the exploration of the structure of materials, to the high-energy accelerators and colliders used to explore the fundamental constituents of matter. A very different regime of very-low-energy, nonneutral plasmas, namely the confinement of nonneutral plasmas in Malmberg-Penning and Paul traps, is yet another modern beneficiary of plasma physics research and has led to an exploration of the fascinating physics regime of crystallized plasma.

- *High-energy-density plasmas.* With the advent of high-power laser facilities, such as those at the University of Rochester, the Naval Research Laboratory, and (under construction) Lawrence Livermore National Laboratory, it becomes possible to explore entirely new regimes of plasma physics in which momentum and energy exchange between photons and matter dominates the physics. This regime of radiation hydrodynamics, which plays important roles in areas such as astrophysics, is experimentally largely unexplored (see Figure 4.1).
- *Nonideal plasmas.* In such plasmas, sometimes also referred to as strongly coupled plasmas, n-body interactions ($n > 2$) become competitive with two-body interactions. Such plasmas arise in the context of high-energy-density matter, which occurs in high-power laser experiments as well as in nature (for example, on the surface of neutron stars).
- *Relativistic plasmas.* The dynamics of relativistic plasmas, as occur in astrophysical jets and in the magnetospheres of compact stellar objects such as neutron stars, as well as in laser-driven plasmas, is largely unexplored by experiment (see Figures 4.1 and 4.2).
- *Dusty plasmas.* Such plasmas arise in a wide range of physical circumstances, including planetary rings and disks, the atmospheres of cool stars, and the interstellar medium, as well as in terrestrial circumstances.
- *Extremely weakly coupled plasmas.* In some physical circumstances, such as accretion onto black holes, there is evidence that coupling between electrons and ions (mainly protons) may be virtually entirely suppressed; in such "advection-dominated accretion flows" (ADAFs), accretion occurs without the usually expected radiative emissions. The plasma physics of such flows is not well understood. It would constitute a major advance if the appropriate physical conditions could be experimentally attained and the physics in this regime explored.
- *Multiphysics/multiresolution simulations.* From the computational perspective, plasma physics is unique in that while the complex (plasma) physical processes do span a large dynamic range in both time and space, these ranges are not so large as to be unattainable by present or future large-scale computations. For this reason, plasma experiments, in both the kinetic and fluid regimes, may provide outstanding validation testbeds for developing kinetic and fluid codes for other disciplines, including astrophysics and space physics.

Another new area of interest for experimental (fusion) plasma research is its connection with astrophysics as fusion experiments directly relevant to important astrophysics problems become more sophisticated and as theory and simulations capable of connecting experiments to astrophysical observations become more sophisticated. The principal areas of interest are the origins of magnetic fields (the dynamo problem) and the dissipation and reconnection of magnetic field lines. To a significant extent, existing fusion experiments (such as tokamaks and reversed-field pinch devices) have been used to study both dynamo action (especially the alpha effect) and reconnection. In addition, there has been a new effort to construct more specialized experiments to explore these issues. Examples include the following:

- Liquid metal dynamo experiments, which try to answer fundamental questions about "slow" and "fast" magnetic dynamos, including nonlinear limiting of the alpha effect and the effects of nonlinearities on turbulent diffusion, and
- Reconnection experiments, ranging from the low-β to the high-β limits and from the collisionless to the fully collisional limits, which try to answer the question of whether (and how) "fast" reconnection can take place and explore the transition from single-fluid MHD to the kinetic regime (see Figures 4.3 and 4.4).

FIGURE 4.1 These two images show two views of the famous Crab Nebula, a remnant of a supernova seen in A.D. 1054. The optical image (top left) shows the highly filamented state of the remnant's gaseous interior that is the result of synchrotron radiation from energetic electrons spiraling in the magnetic field of the remnant. The Chandra X-ray Observatory image of the Crab Nebula (bottom right) shows, for the first time, the x-ray inner ring within the x-ray torus, with the suggestion of a hollow-tube structure for the torus and x-ray knots along the inner ring and (perhaps) along the inward extension of the x-ray jet. Courtesy of (top left) P. Scowen and J. Hester (Arizona State University) and Mt. Palomar Observatories and (bottom right) NASA, Chandra X-ray Observatory Center, and Smithsonian Astrophysical Observatory (SAO).

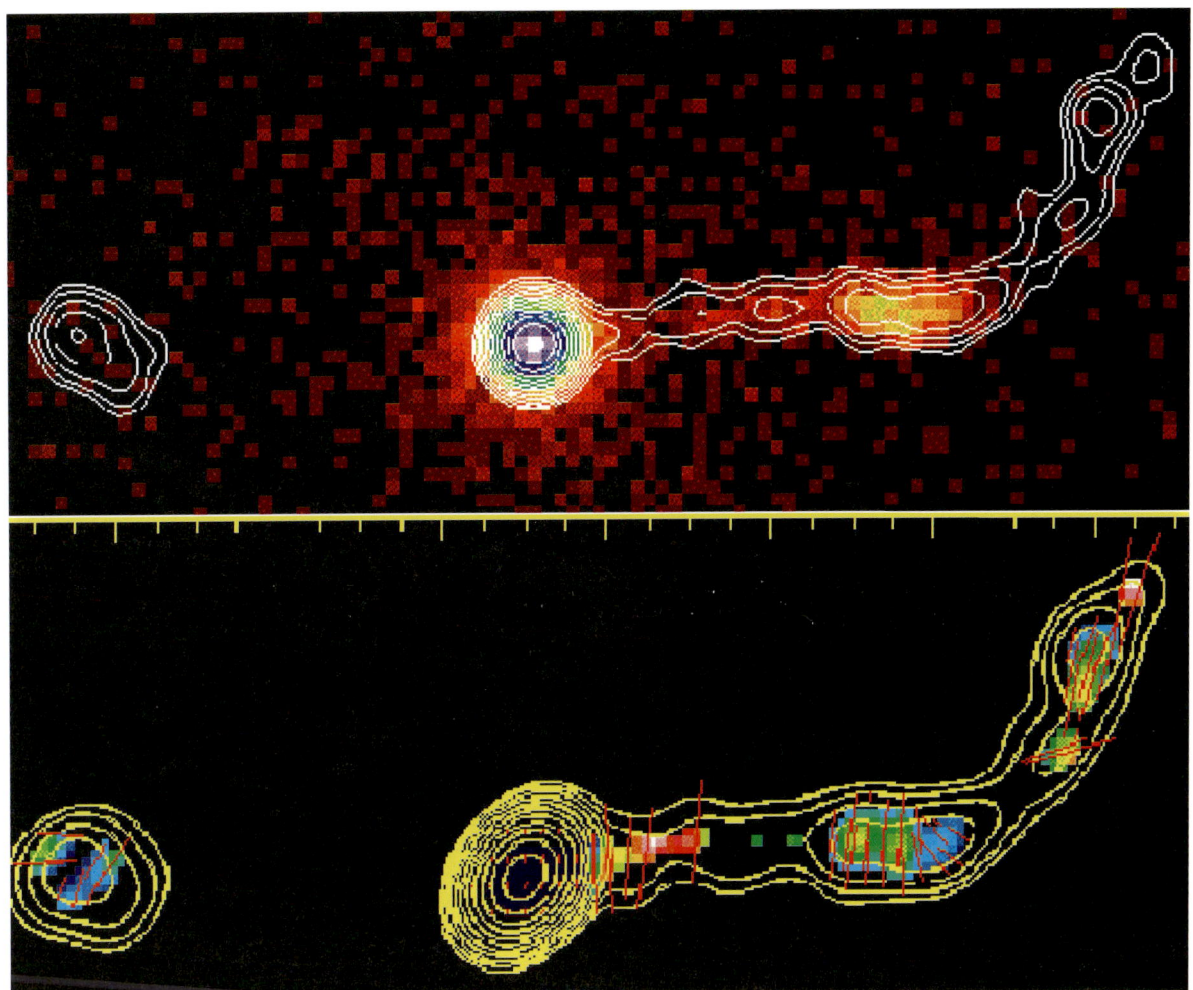

FIGURE 4.2 A Chandra X-ray Observatory (ACIS-S) image (top) of x rays from the quasar PKS 0637-752. The contours show an overlay of the 8.6-GHz emission measured with the ATCA. One tic mark is 1 arcsec, which corresponds to 9.2 kpc in the plane of the sky for the redshift 0.652 of this object. The x-ray jet, which closely correlates with the radio emission from 4 to 10 arcsec west of the nucleus, is the largest and most luminous detected to date. The radio polarization fraction and E-vector direction are shown in the bottom panel. Clearly, major events are taking place 11 arcsec west of the quasar, where the radio jets bend, the x-ray emission drops, and the radio emission becomes unpolarized, followed by realignment of the magnetic field direction. Courtesy of NASA, SAO, D. Schwartz (SAO), and the PKS 0637-752 Consortium.

FIGURE 4.3 This solar image was taken with the TRACE telescope, using a normal incidence soft x-ray mirror whose band pass is centered on an emission line of Fe IX; the image clearly shows the highly complex structuring caused by solar surface magnetic fields and the remarkable "plasma loops," which indicate million-degree solar gases trapped by these magnetic fields. Courtesy of NASA and the Stanford-Lockheed Institute for Space Research.

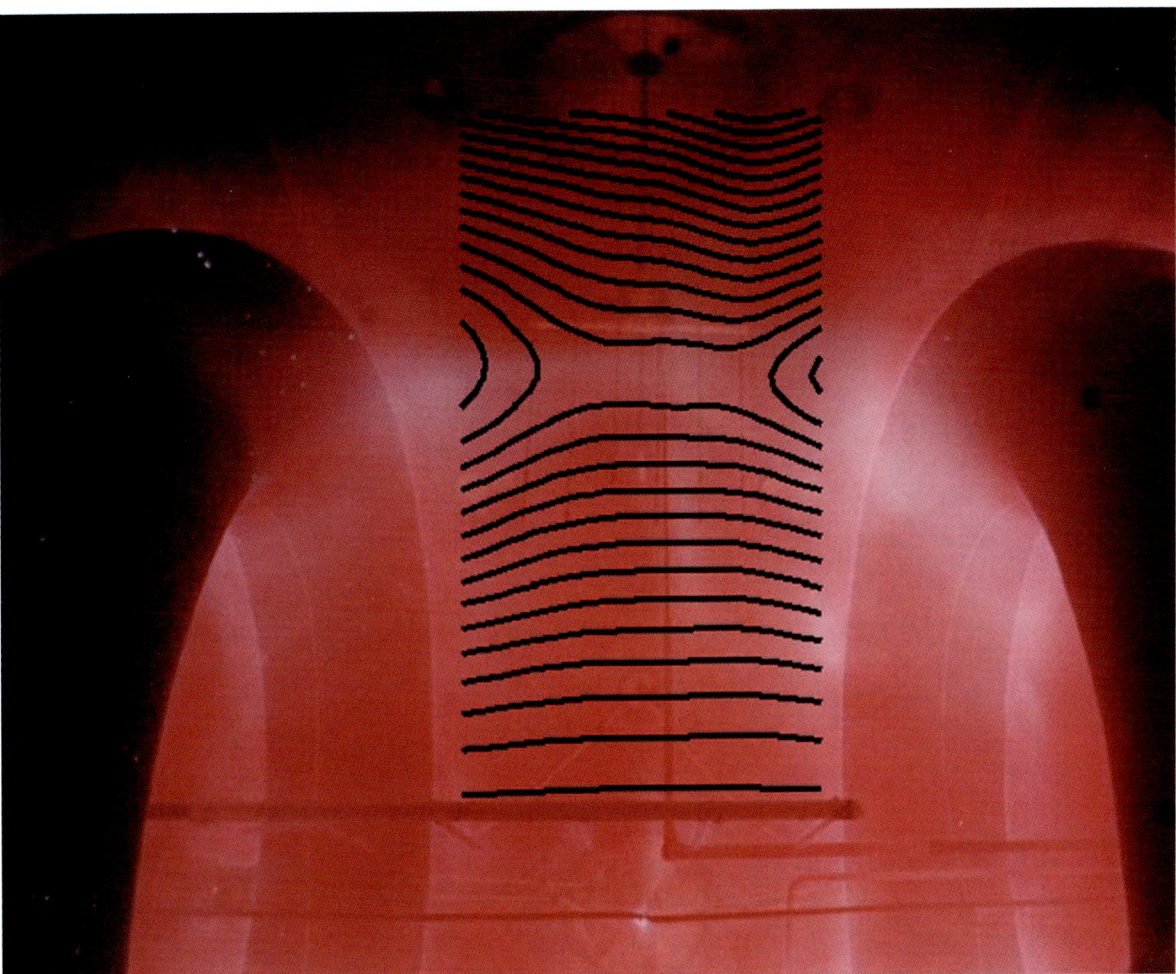

FIGURE 4.4 This image shows the interior of the MRX reconnection experiment. The pair of large coils produce oppositely directed magnetic field lines. The bright area between the coils (in an X-shaped pattern) is the region where magnetic energy is being released into the plasma through the reconnection process. Superimposed is a snapshot of the measured magnetic field. Courtesy of M. Yamada (PPPL). Reprinted by permission from *Journal of Geophysical Research*, 1999, vol. 104, p. 14529.

DOES THE FIELD MAINTAIN LEADERSHIP IN KEY SUPPORTING RESEARCH AREAS?

Quite aside from spawning new ideas that gain currency in other allied fields of physics, fusion research also involves the use and development of tools that it shares with other physical science disciplines. An important question is, To what extent has fusion research maintained the links required to maintain the flow of scientific information between it and allied disciplines? To illustrate the issues involved, the committee has focused on three areas—computational physics, applied mathematics, and experimental tools and techniques for producing and diagnosing plasmas—as examples of how these interactions have evolved.

Computational Physics

Computations are playing an increasingly important role in nearly all areas of science. The effective use of computers requires a close interplay between physical reasoning, approximation, applied mathematics (including algorithm design), and computer science. To ensure that its computational tools remain at the forefront of current technology, a physical discipline must maintain strong interdisciplinary connections to applied mathematics and computer science.

In fusion research, computations have from the early days of the program been an essential tool to understand the highly nonlinear processes characteristic of high-temperature plasmas. Indeed, fusion science played a central role in pushing the development of computations as an ancillary tool for theoretical physics. Thus, fusion research was clearly identified with the frontiers of computational science from its beginnings, the early work on kinetic codes being an outstanding example.

The present fusion program is facing a number of grand challenge computational problems, including the dynamics of short-scale turbulence, which controls energy, particle, and momentum transport in fusion confinement systems, and the evolution of larger scale instabilities, which lead to the rapid loss of magnetic containment on a large scale (disruptive events). Both areas of research are characterized by strong dynamical nonlinearity and a very large range of space- and timescales. In addition to the grand challenge problems, codes have been developed to explore plasma equilibrium and stability in complex geometries; radio-frequency wave propagation and dissipation for exploring heating; and current drive and real-time analysis of data from experiments. Other examples of areas where computations have played a crucial role were discussed in greater depth in Chapter 2.

A fundamental question is whether scientists in the fusion program have developed the computational tools required to explore the complex nonlinear dynamical processes that underlie plasma behavior. Some of these issues were discussed in Chapter 2. Here, the committee briefly discusses two of the grand challenge topics, turbulence and macroscale dynamics, commenting on the computational techniques employed and areas where greater resources or more effort is required.

Scientists studying transport have developed novel magnetic-flux-tube-based coordinate systems for dealing with the extreme anisotropies in the turbulence that develops along and across the magnetic fields confining the hot plasma. The gyrokinetic and gyrofluid techniques, which average over the rapid ion and electron gyromotion, have been developed to uncover the slowly growing instabilities that characterize the fluctuations that occur as a result of the pressure gradients in magnetically confined plasma. These turbulence codes have pushed the frontiers of modern computation in terms of size and complexity and challenges for load-balancing and multiresolution calculations; the sophistication of the physics characterizing these codes is comparable to that of any of the forefront computational areas in modern physics. Visualization techniques developed for understanding the dynamics of large volumes of data in complex geometries are also pushing the state of the art.

In addressing the dynamics of instabilities at scales up to the system size, codes have been developed to effectively treat the complex geometries of modern fusion containment experiments, which is important since the shape of magnetic surfaces can strongly affect the stability and dynamics of large-scale instabilities. In addition, implicit time-stepping techniques have been developed to allow the long-timescale evolution of slowly growing dissipative instabilities. The scientists in the fusion program have been less effective in developing advanced techniques for handling the strong nonlinearities that can develop in plasma systems (Godunov, Flux Corrected Transport (FCT), and related approaches), nor have they been at the forefront of the development of adaptive mesh and similar techniques for addressing the extreme range of scales that is an intrinsic aspect of the dynamics of high-temperature plasmas with very weak dissipation. Stronger links between the fusion community, the astrophysical and space physics community, and the fluid dynamics community would benefit all of these communities. The macroscopic simulation codes are now being ported to the parallel architectures of the most powerful computing platforms, but this task should receive higher priority, because advanced computation in this area requires the effective use of these machines.

Finally, as discussed in Chapter 2, the description of the macroscopic dynamics of high-temperature fusion plasmas requires the use of non-MHD physics at small spatial scales. The inclusion of the requisite kinetic physics and the scale-lengths required to model this physics while at the same time describing the macroscales is a major computational challenge for the program. The development, for example, of adaptive mesh techniques for including this physics would significantly strengthen the physics basis of large-scale modeling of fusion systems and would probably have an impact on the space and astrophysics communities.

Computation is clearly an important force in advancing the understanding of fusion plasmas. In recent years the rapidly increasing ability of computational physicists to successfully model the phenomena observed in experiments has led to a greatly increased demand for active simulation of ongoing experiments. As a consequence, the computational challenges facing the field exceed the resources available, and not enough young scientists are entering the field. The insufficient support also translates into a relatively weak effort to develop new codes that use modern object-oriented programming techniques and into inadequate computer facilities. The latter issue is addressed more fully in the next section.

The DOE has recognized that the lack of support for computation and associated theory is hindering the ability of the program to interpret and understand the plasma dynamics being measured in experiments. In response, it has initiated the Plasma Science Advanced Computation Initiative, which directs new resources into selected areas of computation of importance to the program. This is an essential step toward rectifying a significant deficiency in the program and is strongly endorsed by the committee. The creation of several centers of excellence in plasma science, as recommended by the committee, will also facilitate increased funding in this important area.

Access to State-of-the-Art Computational Platforms

The most powerful computational platform available to scientists in the fusion program is the CRAY T3-E at the National Energy Research Scientific Computing Center (NERSC). This machine is not adequate for the frontier computations in transport (which require the treatment of the cross coupling of electron and ion scales, as discussed in Chapter 2) or macroscopic dynamics (which also requires the treatment of both macro- and microscales). In January 2001, a new IBM-SP, the "Glenn Seaborg" machine (Phase 2, with 2048 processors), became available at NERSC. The new machine, which is comparable to the machines now in use by the DOE DP laboratories, is a significant improvement over the CRAY T3-E. However, the DOE DP laboratories are in the process of procuring even more powerful machines, which will not be accessible to fusion researchers. To develop the predictive capability required for performance predictions in present and future machines, the fusion program will need to keep pace with advances in computational power.

Community Codes

The development of community codes is an essential product of large scientific programs such as fusion since such codes facilitate the exploration of physics issues by all elements of the community, even those that do not have the resources needed to develop code of the complexity that characterizes most plasma problems.

Community equilibrium and stability codes are widely available and shared. The TOQ is a two-dimensional equilibrium code, and VMEC can do both two- and three-dimensional equilibria. MHD stability codes that are widely available include PEST and ERATO/GATO. Kinetic stability codes include FULL and GS2. All of these stability codes can handle the complex shapes that characterize modern plasma confinement machines.

A second class of codes analyzes data from experiments to reconstruct equilibria and analyze energy and particle transport. Examples include TRANSP and EFIT.

Very few of these codes are available on a Web site with appropriate documentation so that they can be simply downloaded and used. Web-based, open-source access to the standard tools of plasma science should be developed by the fusion program to facilitate progress in the field. In many cases, the codes could be upgraded using modern programming to facilitate their future enhancements.

In the last several years there have been several cross-institutional efforts to develop code, including the following:

- *NIMROD, a multi-institutional team project to develop a nonideal MHD code.* The goal of NIMROD is to exploit modern computer methods to model the large-scale dynamics of plasma confinement systems. A similar effort (MH3D) is centered at the Princeton Plasma Physics Laboratory (PPPL). The funding of both efforts has been enhanced through PSACI.
- *The National Transport Code Collaboration (NTCC), a DOE-promoted effort to construct a modular transport code for the community.* The idea behind the modules is to allow disparate (largely Fortran 90) codes to be interfaced, much in the spirit of modern modular codes.

Applied Mathematics

There has traditionally been a strong link between mathematics, especially applied mathematics, and fusion research. In the past, this link was very strong in areas such as stochasticity and chaotic dynamics; multiscale analysis, asymptotics, and boundary layer theory; and weakly nonlinear dynamics. In recent years, these links have developed in the context of problems without any obvious direct application to fusion energy science, so they fall into the category of work on general plasma physics, which has been relatively poorly supported by the DOE fusion program. The funding for the NSF/DOE program in plasma science, now in its fourth year, supports research in basic plasma science and is a positive development, but the overall funding for this program remains small (see Appendix B for figures).

The early links between plasma science and mathematics, which led to significant advances in nonlinear physics (as discussed earlier in this section), are not being maintained despite opportunities for cross-fertilization. Two examples are cited here. First, there has been considerable mathematical interest in the topological structure of magnetic fields. In the fusion context, this interest derived from studies of the confinement properties of magnetic configurations. More recently, it has been realized that field topology is connected to the relaxation and reconnection of magnetic fields, a topic of great importance to fusion science. The sense that fusion researchers are disconnected from this sort of work is reinforced by the fact that important conferences on topological fluid dynamics have not had many participants from the fusion community. A second example is the explosion of work on singularities in continuum systems—for example, How do singularities form in systems described by partial differential equations with smooth initial data? Such problems are of considerable current interest in applied mathematics, fluid dynamics, and astrophysics, but the interaction of these fields with fusion research, which has related problems, is again notable by its absence.

These two examples are emblematic of the larger problem—namely, that fusion science has become disconnected from many of the advances and activities in modern applied mathematics. A number of areas in mathematics have seen dramatic advances over the past few decades, ranging from modern differential geometry and topology to research in the existence and stability of solutions to partial differential equations. Many of these advances have direct applicability to plasma physics. However, the evidence (based on, for example, the topics listed for discussion at recent meetings of the American Physical Society's Division of Plasma Physics) shows that the interaction of plasma science practitioners with these advances is weak and largely confined to practitioners who are increasingly peripheral to the core fusion research program.

Experimental Techniques

The fusion program has developed techniques to produce plasmas and to diagnose their properties. Each of these two endeavors has spawned new tools needed for high-temperature plasmas.

Production Techniques

Using tools developed in the fusion program, plasmas can be produced with temperatures from 10^4 to 10^8 K and densities from about 10^9 to 10^{15} particles/cm^3. Moreover, techniques have been developed to adjust the spatial structure of the plasma—the magnetic field, the plasma temperature, and density profiles. Even aspects of the velocity distribution of particles are adjustable. These techniques have enabled a new area of research—the study of high-temperature plasmas. High-power heating techniques have been developed to deposit megawatts of power into the plasma. Intense beams of neutral atoms (up to about 500 keV in energy) have been developed for this purpose. A wide array of techniques for electromagnetic wave injection has been invented for heating; these employ a huge array of plasma waves spanning the frequency range from tens of kilohertz (where Alfvén waves propagate) to 100 GHz (the electron cyclotron frequency range). A wide assortment of electromagnetic wave sources has been employed or developed (such as high-frequency gyrotrons), as well as a wide assortment of antennas and launching structures. These techniques have also been used to drive current in the plasma and to adjust the current and magnetic field spatial structure. Mega-amperes of plasma current can be driven by these techniques. To fuel the plasma, pellet injectors have been developed that shoot frozen hydrogen pellets into the plasma at high speed. This assortment of plasma production tools has application in many areas, including electromagnetic wave sources, accelerator physics, plasma processing, and beam physics.

Diagnostic Tools

The challenge in diagnosing hot plasmas is to remotely or nonperturbatively measure a large number of fundamental plasma properties. This has required the invention of new methods as well as the extension of existing techniques to new regimes. Techniques using laser injection into plasmas exploit the plasma effect on the laser scattering, the phase angle, and the electric field polarization. Laser scattering has been developed to determine the electron temperature, electron density, and collective density fluctuations. Laser interferometry is used to determine electron density. Measurement of the Faraday rotation of the laser beam yields information on magnetic field. A wide array of spectroscopic techniques, from the visible to the x-ray spectrum, are used to diagnose impurity ion dynamics. The injection of beams of neutral atoms forms the basis for an assortment of measurements, many recently developed. Emission of light from the beam atoms provides a measure of electron density fluctuations. Through a charge exchange interaction between the beam atoms and impurity ions, the impurity ion emission is enhanced, providing a spatially localized diagnosis of impurity ions. The Stark splitting of emission lines from the neutral atoms (produced by the motional electric field experienced by the atoms moving through the magnetic field) yields a measure of the magnetic field in the plasma. Scattering of the beam atoms by protons in the plasma (a Rutherford scattering process) yields information on the temperature and velocity of the protons. Many of these and other diagnostics have been employed to measure detailed aspects of plasma turbulence, as well as equilibrium plasma quantities. The development of novel plasma diagnostics to measure new quantities continues as a vital activity.

HAS THE FIELD BEEN RECOGNIZED FOR ASKING AND ANSWERING DEEP PHYSICS QUESTIONS?

It is the committee's subjective impression that the fusion science field has not received adequate credit for past contributions to other scientific fields and, in spite of doing quality science (as described in this report), is not widely respected as an area of fundamental physics. There is, moreover, evidence (below and in the preceding section) that an inadequate level of formal and informal interaction is taking place between fusion scientists and the rest of the scientific community. The committee supports its conclusion with the following additional data:

• Of the 2185 members of the National Academy of Science, 10 members have worked in fusion research; no one in the field has been elected to the Academy for 13 years. This statistic probably reflects the fact that while the early accomplishments of plasma science received widespread recognition, the increasing isolation of fusion science from the rest of the science community has prevented similar recognition of the more recent accomplishments in the field.
• The National Medal of Science has been awarded to 110 recipients in the physical sciences. Two of the recipients—L. Spitzer and M. Rosenbluth—were specifically rewarded for their fusion-related work. Three recipients—S. Buchsbaum, M. Kruskal, and E. Teller—had worked in plasma physics or fusion at some time in their careers.
• Between 1960 and 1998, there were 180 winners of the Department of Energy's E.O. Lawrence Award. Fifteen of the awards were fusion-related: 10 for magnetic fusion and 5 for inertial fusion.
• The American Institute of Physics has published *Physics News Update* for the past decade. From 1990 through May 2000, the number of physics-related headline items was 1627. Of these, seven (0.4 percent of the total) concerned magnetic fusion, two (0.1 percent) concerned inertial fusion, and four (0.2 percent) concerned general plasma physics.

ARE PLASMA SCIENTISTS WELL REPRESENTED AND TRAINED AT THE NATION'S MAJOR RESEARCH UNIVERSITIES?

Essential to the health of any scientific field is the training and influx of high-quality students on a continuing basis. The major research universities are essential to the maintenance of this process and therefore to the long-term health of the field. There are disturbing demographic trends in plasma science at the major universities that raise significant doubts about the ability of the field to sustain itself over the long term.

Information was provided to the committee from an independent survey[1] of the physics and applied physics faculties at 25 top U.S. universities; the survey indicates that 4 percent of the faculty members are in plasma physics, a number that can be compared with the 6 percent of all members of the American Physical Society who belong to its Division of Plasma Physics. Admittedly, the survey could have been usefully expanded to include more universities and departments, since plasma scientists are also to be found in astrophysics, applied mathematics, and various engineering departments. The disturbing result from this survey is, however, that at these 25 universities, out of roughly 1300 physics faculty, there are only three assistant professors in plasma physics, well below the level at which the plasma physics faculty at these institutions can be sustained. These three assistant professors in plasma physics account for less than 2 percent of the assistant professors in the physics departments at these institutions. This could herald the eventual disappearance of plasma physics from physics departments at the nation's top universities.

According to the American Institute of Physics' *1997 Graduate Student Report* (admittedly already somewhat out of date), 4 percent of graduate students with three or more years of study at Ph.D.-granting physics departments were enrolled in the subfield of plasma and fusion physics, consistent with the fraction of the physics faculty in this subfield. Slightly more recent numbers from the National Research Council on physics and astronomy doctorates by subfield show that 55 out of a total of 1584—or 3.5 percent—were in plasma and high-temperature physics in 1998. Furthermore, students who now graduate with Ph.D. degrees in this subfield are much less likely than before to enter a career in fusion energy science. To illustrate this, about 80 percent of the plasma theory Ph.D. students who graduated during the 1980s from the University of Texas (a major center for fusion and plasma research in the United States) went into fusion positions, whereas only 20 percent of those who graduated during the 1990s did so. This suggests that the pipeline for staffing future plasma physics research is not being filled.

One likely consequence of these demographic shifts is that the fusion field will be increasingly underrepresented at universities, with the demographic balance of Ph.D. scientists shifting to the national laboratories. High-quality fusion science certainly has been and is being done at the national laboratories, and it should be borne in mind that good science can be done in large projects as well as in smaller, university-scale projects. Furthermore, there are many examples of collaborations among fusion scientists at national laboratories and universities, and more such interactions could be encouraged. Nevertheless, the survival of plasma science as a viable field is dependent on maintaining its strength in the university environment.

CREATION OF CENTERS OF EXCELLENCE IN PLASMA AND FUSION SCIENCE

The relatively small number of assistant professors at the large research universities forces the committee to conclude that the field of plasma science is held in low esteem in the academic community at large, an opinion that ripples out to influence students, news media, and policymakers. Given the committee's view that the science coming out of the program is of the highest quality, this is a disturbing conclusion and one for which remedies must be devised and implemented.

While the budget cuts in the program that occurred in FY96 may partially explain the trend, the committee feels that there are deeper problems in the program. Two issues have emerged: (1) the relative isolation of the field—scientists outside the program are apparently not aware of the level of scientific discovery in the program—and (2) the meager degree to which science issues are driving the

---

[1] Survey study conducted by Kenneth Gentle, University of Texas at Austin.

programmatic progress toward fusion. The lack of recognition on the part of outside scientists is reflected in the relatively small number of accolades received by plasma scientists, as discussed previously.

The effective communication of the intrinsically interesting and broadly important science being uncovered in the program is essential if the program is to successfully refurbish its image with scientists and policymakers outside the field. Communication must be enhanced on a broad front. The DOE is actively supporting a distinguished speakers program through its Division of Plasma Physics, a positive but very small step. A fundamental issue, discussed extensively in Chapter 3, concerns the organization of experimental concepts in terms of their progress toward a reactor rather than their progress in addressing important scientific goals. Given the tendency of the program to emphasize performance issues, it is perhaps not surprising that the external community has not fully appreciated the scientific accomplishments of the program. The reorganization of the program around scientific issues would more effectively communicate the science in the program and enhance the ability of the program to address the cross-cutting science issues more effectively.

One way of generating interest and excitement in a field is to offer funding opportunities that will attract the best talent. In a flat funding situation, the maintenance of a continuing flow of talent into the field will be challenging but nevertheless essential to the health of the field—the challenge is to provide appropriate security for productive scientists while at the same time attracting individuals with new ideas and enthusiasm into the field. The committee concludes that the main research groups have remained far too static for the health of the field. To illustrate this point, the list of universities that have major theory groups has not changed in more than 20 years: the University of Maryland, New York University, the Massachusetts Institute of Technology, the University of Wisconsin, the University of Texas at Austin, the University of California at Los Angeles, the University of California at San Diego (UCSD), Cornell University, and the University of California at Berkeley. The strength of the plasma physics programs at Berkeley and Cornell has gone down slightly, while that of the program at UCSD has grown. While the work at all these places is excellent, the committee is concerned that new groups have not spontaneously formed and grown with time. The formation of strong new groups, while not the sole measure of excellence, certainly reflects the success of a field in attracting talent and in making the case that it is ripe for identifying and exploring new science issues.

To address a number of the science issues discussed in Chapters 2 and 3 and the programmatic issues discussed in this chapter, the committee recommends the creation of several interdisciplinary centers (see below). On the scientific side, many of the issues in the fusion program are now sufficiently complex that they require closely interacting, critical-mass groups of scientists to make progress. For example, understanding the dynamics of plasma turbulence and transport requires the development of appropriate physical models; computational algorithms for treating disparate space- and timescales, as well as complex magnetic geometries; efficient programming on massively parallel computing platforms; and an understanding of nonlinear physics (energy cascades, intermittency, phase transitions, avalanches). Tight coupling with a parallel experimental effort is required to challenge theoretical predictions.

No single scientist or small group of practicing scientists has the breadth of knowledge required to tackle such large and complex problems. In the area of theory and computation, the absence of closely interacting teams of critical mass is inhibiting a concerted attack on a number of central science issues confronting the fusion research program. The loose collaborations established by the program from time to time have not succeeded in fostering the close working relationships required to address the most challenging topics.

New "centers of excellence," each of which would either serve as the node for a distributed network of collaborators or undertake scientific explorations of significant magnitude at one site or would do both, could create a new focus on scientific issues within the U.S. fusion program. (See discussion and recommendations in Chapters 2 and 3.)

Potential focus topics for such centers include turbulence and transport, magnetic reconnection, energetic particle dynamics, and materials; other topics would emerge in a widely advertised proposal process. Because the topics listed above have such broad scientific applicability in allied fields, collaborations could be set up with scientists having expertise of great value to the plasma science and fusion research effort. An explicit goal of the centers should be to convey important scientific results to both the fusion community and the broader scientific community. The mere announcement of opportunity for fusion centers of excellence would signal to the broader scientific community the fusion community's intent to significantly bolster the scientific strength of the field.

FINDINGS AND RECOMMENDATIONS

Findings

1. *In the key scientific areas being discussed, there is a history of intellectual exchange between the plasma physics community and the broader scientific community in areas such as MHD, nonlinear dynamics, instabilities, and transport.*

2. *The current fusion program is relatively weakly coupled to programs in the rest of the physics community.* Most of its external coupling takes place in areas such as space physics and (plasma) astrophysics, which are themselves poorly represented in physics departments. Believing that the education of all graduate students in physics should include a course in plasma physics, the committee is distressed by the small percentage (roughly half) of departments with faculty in plasma physics.

3. *The future representation of plasma science at universities is threatened by an apparent lack of new blood.* Of a total physics faculty of roughly 1300 individuals at the top 25 physics research universities, there are only 3 assistant professors of plasma physics. The committee is concerned that the very low overall rate of replacement of plasma physics faculty threatens the future of this field.

4. *In the area of theory and computation, the dearth of critical-mass, closely interacting teams of researchers from different institutions is inhibiting a concerted attack on a number of important science issues faced by the fusion research program.* The loose collaborations established from time to time by the program have generally not succeeded in nurturing the close working relationships required to address the most challenging topics. No single scientist or small group of practicing scientists has the breadth of knowledge required to tackle such large and complex problems.

Recommendations

A systematic effort to reduce the scientific isolation of the fusion research community from the rest of the scientific community is urgently needed.

Program planning, funding, and administration should all encourage connectivity with the broad scientific community. The community of fusion scientists should make a special effort to communicate its concepts, methods, tools, and results to the wider world of science, which is largely unaware of that community's recent scientific accomplishments.

There are numerous examples in federally funded research programs of formal coordination mechanisms having been established among related programs in different agencies. In some instances this coordination can optimize the use of funding. Perhaps more significant, dialogue among the leaders of these government research programs can encourage interactions among the various scientific communities, foster joint undertakings, and raise the visibility of the discipline as a whole.

The fusion science program should be broadened both in terms of its institutional base and its reach into the wider scientific community; it should also be open to evolution in its content and structure as it strengthens its research portfolio.

Clearly, this issue can be approached in a number of ways. One good way would be to set up competitive funding opportunities of sufficient magnitude to elicit responses from potential new institutional participants. The creation of centers of excellence in fusion science (proposed below) and the greater involvement of the National Science Foundation in fusion and plasma science are other ways to broaden the institutional base of fusion science.

A larger proportion of fusion funding could be made available through open, well-advertised, competitive, peer-reviewed solicitations for proposals. Fusion program peer review could involve scientists from outside the fusion community where appropriate. The evaluation and ranking of proposals by panels that include individuals with appropriate expertise in allied fields would broaden the intellectual reach of the grant review process.

Plasma science research not immediately related to the quest for a practical source of power from fusion (the fusion energy goal) should play a more influential role in the fusion program, at an appropriate budget level. This would include funding for general plasma science and for peer-reviewed individual investigator grants, each presently a small fraction of the total fusion energy science program (see Appendix B). Such funding would encourage new interchanges that enrich fusion science.

To ensure that increasing institutional diversity is a continuing goal, the committee recommends that the breadth and flexibility of participation in the fusion energy science program should be a program metric.

Several new centers, selected through a competitive peer-review process and devoted to exploring the frontiers of fusion science, are needed for both scientific and institutional reasons.

Many of the issues in fusion science are now of sufficient complexity that they require closely interacting, critical-mass groups of scientists to make progress. For example, understanding the dynamics of plasma turbulence and transport requires the development of appropriate physical models; computational algorithms for treating disparate space- and timescales, as well as complex magnetic geometries; efficient programming on massively parallel computing platforms; and an understanding of nonlinear physics (energy cascades, intermittency, phase transitions, avalanches). No single scientist and no small collaboration of practicing scientists has the breadth of knowledge required to tackle such large and complex problems.

The centers of excellence could create a new focus on scientific issues for the U.S. fusion program. A center could serve as a node for a distributed network of close collaborators or it could undertake scientific explorations of significant magnitude, or it could do both. The centers would combine the expertise and approaches of national laboratories on the one hand and universities on the other. They should have a number of programmatic and structural features that will let them play an appropriate role in addressing the critical problems of the field:

- A proposal for a center should have a plan to identify, pose, and answer scientific questions whose importance is widely recognized.

- One size cannot meet all scientific challenges. The committee envisions a center comparable in size to current NSF-sponsored centers (with operating costs of $1 million to $5 million per year), although the size should ultimately be determined by the proposal process. Some centers may need on-site experimental facilities and some may need only computing facilities and access to larger national computer centers.
- A team of between four and six coinvestigators with broad expertise and connections to other research groups and laboratories should form the core of the center's personnel. This team should be augmented by a similar number of temporary research staff (funded, at least in part, by the center) and an appropriate number of support staff.
- The center should enable links to various scientific disciplines, including physics, mathematics, and computer science, depending on the problem it is focusing on. It should have a plan for bringing practitioners of other disciplines, from other institutions, into the fusion community and should make the community's experimental resources more widely available.
- The institutions housing or participating in such centers should make a commitment to add faculty or permanent research staff, as appropriate, in plasma/fusion science and/or related areas.
- The centers should have a strong educational component, featuring outreach to local high schools, undergraduate research opportunities, and a graduate research program.
- Centers should sponsor multidisciplinary workshops and summer schools focused on their central problem that will bring together students and researchers from various fields and institutions. The workshops would aim to bring in new ideas and collaborators as well as to disseminate to other fields the results they are achieving as they address the fundamental problems of fusion science.

Potential focus topics for centers include turbulence and transport, magnetic reconnection, energetic particle dynamics, and materials; other topics would emerge in a widely advertised proposal process. Topics such as these are of broad scientific interest in allied fields. To build another bridge to allied fields, the DOE should cooperate with NSF in establishing one or more centers addressing a topic of general interest in plasma science. The DOE/NSF centers should have as a central objective establishing collaborations with scientists possessing expertise of value to the plasma science and fusion research effort. An explicit goal of the centers should be to convey important scientific results to the broader scientific community as well as the fusion community. Even just an announcement of opportunity for fusion centers of excellence would signal to the broader scientific community the fusion community's intent to significantly bolster the scientific strength of the field.

It would be highly desirable for other federal agencies, particularly the NSF, to collaborate in one or more fusion centers for reasons of disciplinary and institutional diversity as well as to obtain the benefits of interagency collaboration. However, DOE should play a lead role in these centers, not only for reasons of administrative clarity but also because its leadership will ensure that the technical capabilities of the fusion energy science community are made available to new participants. DOE leadership will also ensure that progress at the centers is communicated throughout the fusion community and translated into DOE program plans, to hasten the progress toward the fusion energy goal.

The procedure for awarding grants for fusion centers of excellence could do much to remedy the isolation of the fusion science community by ensuring that the broader scientific community participates in the institution-building effort. The selection process for the centers should feature open, competitive peer review employing clear, science-based selection criteria, as outlined above.

The committee believes this recommendation to be critical enough to the new science-based approach to fusion energy that ways should be found to fund a first center *even in a level budget scenario.* The success of the competition and the quality of the first center should guide the decision whether to launch a second or even a third center. In other programs, such centers have been effective mechanisms for strengthening the breadth and depth of a broad scientific area. In the committee's view, there is a very strong argument for expanding program funding to give fusion centers of excellence a strong and durable foundation.

The National Science Foundation should play a role in extending the reach of fusion science and in sponsoring general plasma science.

The mission of OFES, following the restructuring of the program in 1996, has been to establish the knowledge base in plasma physics required for fusion energy, with the result that a substantial number of plasma science issues are being explored within the fusion regime that also have applicability to allied fields such as astrophysics. For this reason, the committee believes that NSF should begin to play a larger role in the solution of these basic plasma science issues. The involvement of NSF could have an intellectual impact on basic plasma science similar to that which it has had on basic research in other scientific disciplines where mission agencies like DOE play the main funding role. NSF involvement would facilitate linkage to other fields and the involvement of new scientists in the program.

Recently, NSF and DOE collaborated on a small but highly effective program to encourage small-laboratory plasma experiments and the theoretical exploration of topics in general plasma science. The large number of proposals submitted to this program is an indication of the need for it. The rationale for the expansion of research in general plasma science was well articulated in the NRC's *Plasma Science: From Fundamental Research to Technological Applications*, National Academy Press, Washington, D.C., 1995.

The NSF/DOE plasma science initiative, if operated at a dollar level closer to that contemplated in the *Plasma Science* report (an additional $15 million per year for basic experiments in plasma science), can serve more than one important function:

- Stimulating research on broad issues in plasma science that have potential applications to fusion and
- Enhancing interagency cooperation and cultural exchange on the approaches used by the two agencies for defining program opportunities, disseminating information on research results to the scientific community, selecting awardees, and judging the outcomes of grants.

The optimal process for this partnership, if there is sufficient funding (as requested in the *Plasma Science* report), would be an annual solicitation of requests for proposals (RFPs). In particular, this frequency would give new Ph.D.s the chance to enter the field and stay in it, since new Ph.D.s are produced by degree-granting institutions each year and new graduate students enter school each year.

Another limitation of the ongoing NSF/DOE program in basic plasma science is the absence of any provision for modest experiments in the $1 million per year class. Historically, neither DOE nor NSF has funded plasma science experiments of this scale. For this reason, the committee recommends a cooperative NSF/DOE effort to broaden the scientific and institutional reach of fusion and plasma research to obtain valuable scientific results. Increased NSF funding and a stronger focus on fusion as well as plasma science within NSF would be required. As discussed in the preceding recommendation (recommendation 4 in the Executive Summary), NSF could cosponsor one or more centers of excellence in fusion and plasma science.

Appendixes

A

Summary of Committee Meetings

The committee held three data-gathering meetings in 1999 and two mainly deliberative meetings in 2000.

The first meeting, a data-gathering meeting that provided the information for the interim report, was held in La Jolla, California, May 16-19, 1999. The committee heard talks on the OFES program perspective, two talks on the fusion program portfolio, and two talks on the international standing of the U.S. fusion program by respected scientists from India and Japan. Bruno Coppi, of the Massachusetts Institute of Technology, served as the official respondent and spoke after each of the talks. The committee then broke into two groups to hear more detailed, concurrent talks, one on the theoretical component of the program, the other on the experimental research component. The full committee discussed the talks to the two groups and drafted an outline and some text for the interim report. This draft was then reviewed by the steering committee at the end of the meeting.

The second meeting was held in conjunction with the 1999 Fusion Summer Study in Snowmass, Colorado, July 20-23, 1999. The committee spent some time hearing from specific researchers, including one researcher who spoke on ICF/IFE/MFE programmatic and scientific connections, but it spent most of the time observing the plenary sessions of this first-time-ever community process. The nonfusion scientists on the committee particularly found this meeting to be educational. The committee also spent time finishing the interim report and talking with the FESAC chair, John Sheffield.

The third meeting was held in conjunction with the meeting of the American Physical Society's Division of Plasma Physics in Seattle, November 17-18, 1999. The committee heard talks on the perspectives of FESAC, the national laboratories, and part of the university community. There were also talks on the role of fusion energy in U.S. energy policy, the scientific synergism between magnetic and inertial fusion energy, and OFES decision processes. Several researchers spoke on liquid wall technology. The remainder of the committee's meeting was spent in closed deliberations.

The fourth meeting was held in Washington, D.C., February 23-25, 2000. The committee heard presentations on the demographic distribution of the field and related issues. They also heard views from the Office of Management and Budget's Examiner for OFES and the Associate Director for Fusion

Energy Sciences. Next, the committee broke up into working groups, each of which drafted a chapter of the final report. It then came together to review the drafts as a full committee.

The fifth, and final, meeting of the committee was held at the National Academy of Sciences' Beckman Center in Irvine, California, May 8-9, 2000. The committee worked on the draft chapters and finalized the findings and recommendations of the report.

B

Funding Data

TOTAL BUDGET OF THE OFFICE OF FUSION ENERGY SCIENCES

The total OFES budget figures shown in Figure B.1 are the Congressional appropriation numbers in FY00 dollars and do not include general reduction costs to the program. The allocation of OFES funding to particular categories is shown in Table B.1 for FY97 to FY00. The percentage shares for FY00 are shown in Figure B.2.

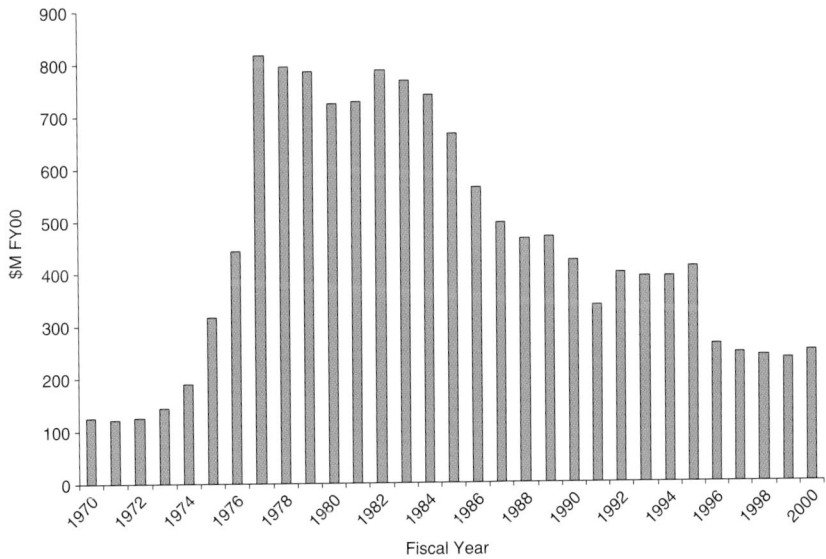

FIGURE B.1 Congressional funding of magnetic fusion R&D, FY70 to FY00. SOURCE: Richard E. Rowberg. 2000. *Congress and the Fusion Energy Sciences Program: A Historical Analysis*, Washington, D.C.: Congressional Research Service.

TABLE B.1 OFES Budget Distribution Since FY97 (million spent dollars)

Budget Category	Fiscal Year			
	1997	1998	1999	2000
Tokamak research	48.7	46.2	45.8	47.1
Facility operations	60.8	56.1	60.2	70.1
Alternative concepts	16.4	24	37.3	52.3
Theory	18.4	19.8	22.7	24.6
General plasma science	3.8	5.1	6.2	8.2
Other	14.3	6.9	6.9	7.4
Materials research	6	7.7	6.8	7.2
Engineering research	56.3	58.3	36.7	27.8
Total	224.7	224.1	222.6	244.7

SOURCE: Office of Fusion Energy Sciences.

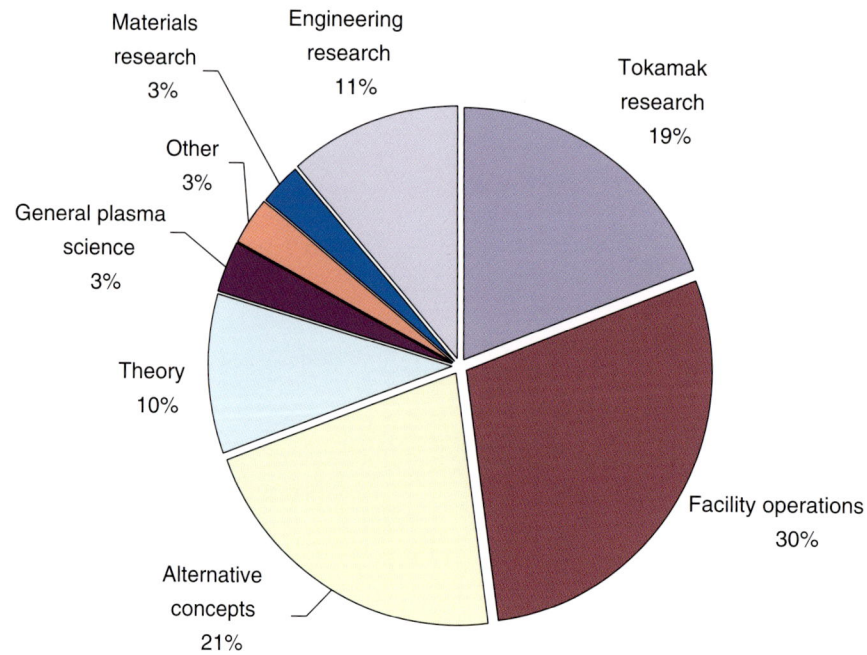

FIGURE B.2 Funding profile for OFES in FY00, $245 million spent dollars.

APPENDIX B

NATIONAL SCIENCE FOUNDATION/DEPARTMENT OF ENERGY PLASMA PHYSICS PARTNERSHIP FUNDING

According to OFES, it spent the following amounts under the NSF/DOE plasma physics partnership: FY97, $2.030 million; FY98, $2.878 million; FY99, $3.369 million; FY00, $4.121 million. According to the NSF, it spent the following amounts for the NSF/DOE plasma physics partnership (these figures do not represent the full NSF investment in plasma science and engineering; instead, they represent contributions for the partnership): FY97, $2.75 million; FY98, $2.75 million; FY99, $2.75 million; and FY00, $3.00 million.

OFFICE OF FUSION ENERGY SCIENCES FUNDING TO UNIVERSITIES

In FY99 and FY00, OFES funding for various university research activities in fusion energy science was distributed as shown in Table B.2 (the total OFES budget was about $240 million): The sharp drop-off in enabling R&D funding between 1999 and 2000 is associated with a decrease in pass-through monies for ITER superconducting magnet development work being done at the Massachusetts Institute of Technology.

Funding of fusion energy sciences at universities by the OFES with and without MIT is shown in Figure B.3. These data do not include pass-through funds sent to universities for ITER-specific expenses that were not spent on research, such as operation of the ITER co-center and industrial tasks (magnets, etc.) in 1992 to 1999. FY00 dollars are actual figures through April 2000 and estimates for the rest of the fiscal year.

Figure B.4 shows the distribution of DOE funding for three categories of R&D at colleges and universities (total of direct and indirect costs). These numbers reflect data reported by DOE to the Office of Management and Budget under Schedule C, categories 1441-01 (direct) and 1442-01(indirect). They exclude funds for FFRDCs administered by universities (e.g., PPPL and the Stanford Linear Accelerator Center (SLAC)). The total for each program is the total appropriated minus any below-the-line general or targeted reductions (e.g., contractor travel reductions).

TABLE B.2 OFES Funding for Fusion Science Research at Universities (million dollars)

Item	FY99	FY00
Major facilities (PPPL, General Atomics, MIT)	19.235	19.744
Smaller facilities	13.604	16.671
Diagnostics	3.232	3.210
Theory	8.551	9.126
General plasma science	4.310	5.103
Enabling R&D	16.553	9.192
Total	65.485	63.046

SOURCE: University Fusion Association.

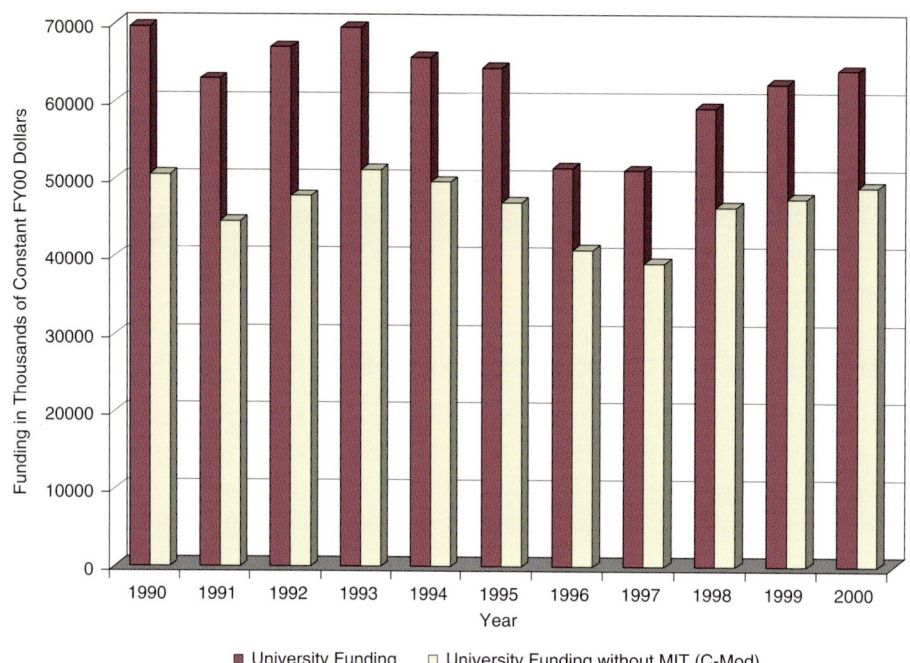

FIGURE B.3 Office of Fusion Sciences funding of fusion energy sciences at universities, with and without the Massachusetts Institute of Technology. SOURCE: Office of Fusion Energy Sciences.

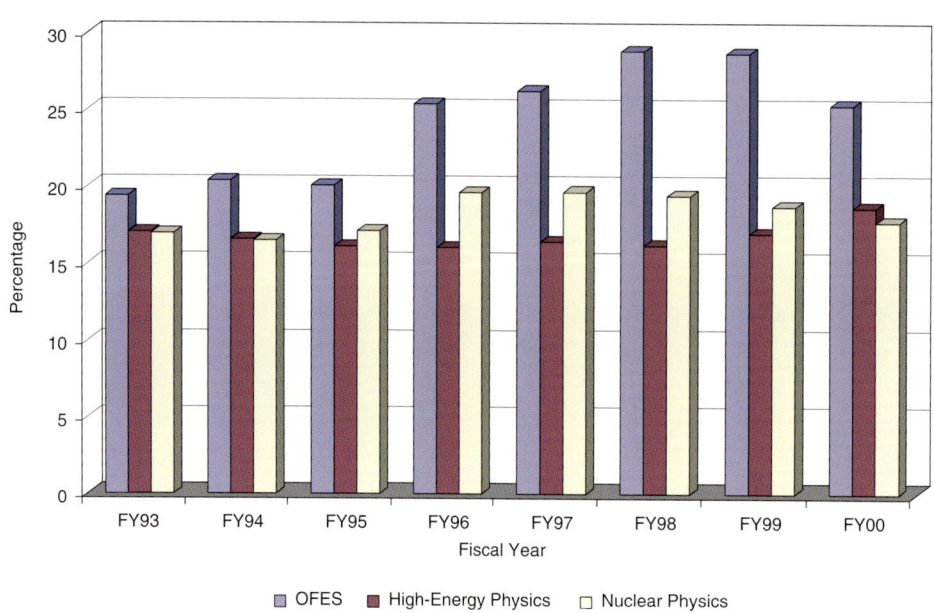

FIGURE B.4 Distribution of university funding for DOE's fusion and high-energy and nuclear physics programs as a percentage of the total funding for those programs. SOURCE: DOE Congressional Funding Office via the Office of Management and Budget.

C

The Family of Magnetic Confinement Configurations

The family of configurations may be characterized at one extreme as being externally controlled and at the other extreme as being self-organized. Externally controlled configurations generally have strong, externally applied magnetic fields, minimizing the need for internal plasma currents and imposing favorable stability. However, such systems tend to be large, with relatively low values of β. Self-organized systems generally employ only a weak, external magnetic field but require strong, internally generated plasma current. The absence of a strong external field tends to reduce the stability and confinement. However, such systems tend to be smaller and simpler. There is a nearly continuous spectrum of configurations, from the most externally controlled (such as the stellarator) to the most self-organized (such as the field-reversed configuration). Externally controlled configurations tend to be the most advanced, and self-organized configurations tend to be at the exploratory stage.

The vocabulary of fusion concepts is rather unfortunate—each variation of the magnetic configuration is labeled with a distinct and rather nondescriptive name. While it provides a useful taxonomy for fusion scientists, it can give the wrong impression—that is, that the family of concepts is a disjointed set of trial-and-error experiments, when it is, in fact, a set of variations that are connected by common plasma physics. The ultimate fusion concept may emerge from the present list of concepts, it may evolve as a hybrid of these concepts, or it may be an as-yet-undiscovered approach.

Elaboration of the features of several representative configurations serves to illustrate the breadth of the family. This description begins at the externally controlled extreme and proceeds toward self-organized systems.

THE STELLARATOR

The stellarator is a configuration in which no currents within the plasma are required for confinement. The stellarator was invented by U.S. scientists but has been most extensively explored by German and, more recently, Japanese scientists. Strong external magnets of very complicated structure produce a three-dimensional plasma equilibrium in which there is no direction of symmetry. Stellarators would result in a relatively large reactor with superconducting coils. However, they are inherently steady-state and are free of the instabilities or disruptions that would be driven by the plasma current. Stellarators offer physics advances by enabling the study of plasma stability in the absence of plasma current and the investigation of new symmetry principles. For example, in the new quasi-symmetric stellarator configuration, a complicated three-dimensional magnetic structure would appear as nearly two-dimensional to an orbiting particle.

THE TOKAMAK

Tokamaks are also externally controlled, but less so than the stellarator. The tokamak was invented by scientists from the former Soviet Union. A strong toroidal magnetic field is applied externally, but plasma current is required to produce a weaker magnetic field, which is directed along the shorter (poloidal) direction. The tokamak is two-dimensional (there is symmetry in the toroidal direction). The most highly studied configuration, the tokamak has contributed enormously to numerous areas of plasma physics, and it serves as an informal standard against which other configurations can be compared in reactor attributes.

THE SPHERICAL TORUS

As one reduces the aspect ratio of the tokamak so that the hole in the center of the torus becomes very small, the β stability limit increases. This relatively compact, high-pressure fusion reactor concept is known as the spherical torus. The configuration can uncover tokamak physics at the geometric extreme of small aspect ratio, where the pressure limit and the pressure-driven self-current (the bootstrap current) are expected to be very large. The virtues of the configuration were extolled by U.S. scientists, but a spherical torus was first successfully built and tested in the United Kingdom.

THE REVERSED-FIELD PINCH

As the externally applied toroidal magnetic field of the tokamak is reduced by a factor of 10, the plasma becomes more self-organized. This configuration, known as the reversed-field pinch (because the toroidal magnetic field reverses direction with radius), offers possible reactor advantages by eliminating the need for a strong toroidal field. However, the weaker magnetic field reduces the stability and confinement of the plasma. The reversed-field pinch provides an experimental vehicle with which to investigate the behavior of magnetic field turbulence and relaxation relevant to a range of natural and fusion plasmas.

THE SPHEROMAK AND THE FIELD-REVERSED CONFIGURATION

At the extreme of self-organized plasmas are toroidal plasmas, which are taken to the limit of unity aspect ratio—the central hole is eliminated. Such a plasma is potentially very attractive as a fusion energy source; it is extremely compact (nearly a sphere) and requires no external magnets. However, the macroscopic stability and confinement of such configurations may be degraded. Two examples of such compact toroids are the spheromak (which contains both poloidal and toroidal magnetic fields generated by plasma currents) and the field-reversed configuration (the simplest geometry, containing only poloidal fields).

OTHER CONCEPTS

There are concepts under study that cannot be categorized in the externally controlled/self-organized scheme. Two examples are magnetized target fusion and the dipole configuration. Magnetized target fusion compresses a compact toroid to densities intermediate between those of magnetically and inertially confined plasmas. It offers a new, perhaps simpler approach to fusion, as well as access to plasma physics regimes. The dipole configuration mimics confinement of plasma in planetary magnetospheres, again offering special advantages and physics insights.

D

Glossary

Alcator: A tokamak device in the United States.
Alfvén waves: A low-frequency, transverse wave (lower in frequency than the gyration frequencies of electrons and ions) in which the plasma and magnetic field move together.
ASDEX: A tokamak device in Germany.
Beta (β): Ratio of the pressure of the plasma to the "pressure" (energy density) of the confining magnetic field.
Charge exchange: Mechanism by which an ion in an overall charge-neutral beam takes the charge from a plasma ion.
DIII-D: A tokamak device in the United States.
ELM: An instability at the edge of a plasma that is operating in the H-mode.
Flux surfaces: Surfaces, usually nested, on which magnetic field lines lie.
FRC: A magnetic confinement configuration where currents generated by the plasma's pressure cause the magnetic field to reverse, resulting in self-confinement.
Fusion-grade: A plasma with a temperature of the order of 100 million degrees, or around 10 keV.
Gyrofluid: A fluid-like description of plasma dynamics constructed from a finite number of velocity moments of the gyrokinetic equations. This construction includes collisionless damping due to resonant particle interactions.
Gyrokinetic: A reduced description of plasma dynamics obtained by averaging out the fast gyrating motion of particles around field lines.
Gyroradius: The radius of gyration of a particle around a magnetic field.
H-mode: Operational regime in which tokamak confinement, even with external heating, is significantly higher than tokamak confinement in L-mode. The H-mode enhances confinement results through the formation of a transport barrier at the plasma edge.

ITG: Instability driven by the gradient of the ion temperature.

JET: A tokamak device in Europe in which deuterium/tritium fusion experiments have been completed.

JT-60U: A tokamak device in Japan.

L-mode: Operational regime in which tokamak confinement is degraded when the plasma is heated.

MHD: A model of a plasma as a fluid in a magnetic field. Ideal MHD is when the fluid is taken to have zero resistivity, in which case the plasma and magnetic field are constrained to move together. This limit maximizes plasma stability.

Motional Stark effect: A plasma diagnostic that measures the polarization of light due to Stark splitting of atomic energy levels.

MRX: The Magnetic Reconnection Experiment is a small laboratory experiment located at the Princeton Plasma Physics Laboratory. The goal of MRX is to investigate the fundamental physics of magnetic field line reconnection, an important process in magnetized plasmas in space and in the laboratory.

Radio-frequency waves: A generic designation for waves in a plasma driven by external electromagnetic power sources and having frequencies anywhere from kilohertz (Alfvén waves) to multigigahertz (electron-cyclotron waves).

Reconnection: A process in which the reversed components of adjacent magnetic field lines cross-connect with each other, releasing magnetic energy into high-velocity flows.

Reversed-field pinch: A magnetic confinement configuration with toroidal symmetry, where the magnetic field in the toroidal direction and that in the transverse field are equally strong and the former reverses direction inside the plasma.

Sawtooth: Periodic instability of the core of tokamaks in which a slow increase in the electron temperature is followed by a fast decline.

Separatrix: A bounding line in magnetic topology across which the direction of a field line reverses.

Spherical torus: Doughnut-shaped magnetic confinement configuration in which the magnetic fields in the toroidal and transverse directions are comparable.

Spheromak: Spherical magnetic confinement configuration in which the magnetic fields in the long and short directions of the doughnut are comparable in strength and the fields are primarily generated by internal plasma currents.

Stellarator: Magnetic confinement configuration in which all of the confining magnetic fields are created by external coils, usually in the form of helical coils wound into a torus.

Tearing mode: Small-amplitude version of magnetic field line reconnection.

TFTR: A tokamak device in the United States in which deuterium/tritium fusion experiments have been carried out.

Tokamak: Doughnut-shaped magnetic confinement configuration in which the magnetic field in the toroidal direction is much stronger than in the transverse direction.

TRACE: The Transition Region and Coronal Explorer is a NASA Small Explorer mission to image the solar corona and transition region at high angular and temporal resolution.

E

Acronyms and Abbreviations

ADAF:	advection-dominated accretion flow
DOE:	Department of Energy
DP:	Defense Programs
EFIT:	a computer modeling code
ELM:	edge-localized mode
ERATO:	a computer modeling code
FCT:	flux-corrected transport
FESAC:	Fusion Energy Sciences Advisory Committee
FIRE:	Fusion Ignition Research Experiment
FRC:	field-reversed configuration
FULL:	a computer modeling code
GA:	General Atomics
GATO:	a computer modeling code
GS2:	a computer modeling code
IFE:	inertial fusion energy
ITER:	International Toroidal Experimental Reactor
JET:	Joint European Torus
KAM:	Kolmogorov-Arnold-Moser
MFE:	magnetic fusion energy
MHD:	magnetohydrodynamics
NASA:	National Aeronautics and Space Administration
NERSC:	National Energy Research Scientific Computing Center
NIMROD:	Non-ideal MHD with Rotation Open Discussion project
NRC:	National Research Council
NSF:	National Science Foundation
OER:	Office of Energy Research (DOE; renamed the Office of Science)

OFES:	Office of Fusion Energy Sciences
OMB:	Office of Management and Budget
PEST:	a computer modeling code
PPPL:	Princeton Plasma Physics Laboratory
PSACI:	Plasma Science Advanced Computation Initiative
SEAB:	Secretary of Energy Advisory Board
TFTR:	Tokamak Fusion Test Reactor
TRANSP:	a computer modeling code
VMEC:	a computer modeling code